人工智能技术丛书

轻松学会
TensorFlow 2.0
人工智能深度学习应用开发

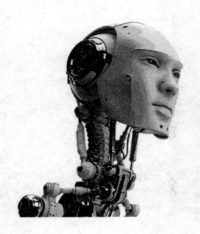

黄士嘉　林邑撰　著

清华大学出版社
北京

内 容 简 介

本书从介绍深度学习和重要入门知识入手，通过范例讲解 TensorFlow 的应用开发。本书文字清晰、严谨，并辅以简洁明了的插图说明，同时提供步骤细致的范例程序教学，让读者可以轻松理解并掌握深度学习原理和 TensorFlow 开发方法。

本书分为 12 章，内容包括：环境安装、TensorFlow 2.0 介绍、回归问题、二分类问题、多分类问题、神经网络训练技巧、TensorFlow 2.0 高级技巧、TensorBoard 高级技巧、卷积神经网络经典架构、迁移学习、变分自编码器和生成式对抗网络。

本书适合 TensorFlow 深度学习自学者、深度学习开发人员、人工智能行业咨询顾问等阅读，也适合作为高等院校和培训学校人工智能及其相关专业师生的教学参考书。

本书为博硕文化股份有限公司授权出版发行的中文简体字版本
北京市版权局著作权合同登记号　图字：01-2020-5817

本书封面贴有清华大学出版社防伪标签，无标签者不得销售。
版权所有，侵权必究。举报：010-62782989，beiqinquan@tup.tsinghua.edu.cn。

图书在版编目（CIP）数据

轻松学会 TensorFlow 2.0 人工智能深度学习应用开发/黄士嘉，林邑撰著.—北京：清华大学出版社，2021.1
（人工智能技术丛书）
ISBN 978-7-302-56645-8

Ⅰ.①轻… Ⅱ.①黄… ②林… Ⅲ.①人工智能—算法—研究 Ⅳ.①TP18

中国版本图书馆 CIP 数据核字（2020）第 202780 号

责任编辑：夏毓彦
封面设计：王　翔
责任校对：闫秀华
责任印制：丛怀宇

出版发行：清华大学出版社
　　　网　　　址：http://www.tup.com.cn，http://www.wqbook.com
　　　地　　　址：北京清华大学学研大厦 A 座　　　邮　　编：100084
　　　社　总　机：010-62770175　　　　　　　　　邮　　购：010-62786544
　　　投稿与读者服务：010-62776969，c-service@tup.tsinghua.edu.cn
　　　质　量　反　馈：010-62772015，zhiliang@tup.tsinghua.edu.cn
印 装 者：北京国马印刷厂
经　　销：全国新华书店
开　　本：190mm×260mm　　　印　张：18.5　　　字　数：474 千字
版　　次：2021 年 1 月第 1 版　　　　　　　　　　印　次：2021 年 1 月第 1 次印刷
定　　价：79.00 元

产品编号：087172-01

改编说明

当今是人工智能的好时代，也是机器学习技术日新月异的时代，更是深度学习给机器带来真正智慧的时代。深度学习是机器学习领域中一个全新的研究方向和应用热点，它是机器学习的一种，也是实现人工智能的必由之路。

深度学习近十年的发展让我们见证了它在对文字、图像、语音、自然语言理解、搜索技术和机器翻译等诸多领域取得的突破性进展和令人惊叹的成功，由此开启了各行各业中各种AI应用无所不在的盛况。单纯看深度学习的研究是一门艰深的科学，让大多数人望而却步，无从下手。然而，这方面的需求却是多方面、多个层次的。例如，开发深度学习的应用并不需要每个人都是人工智能深度学习研究方面的专家。更多想入行的人需要的是一本人工智能深度学习应用开发快速入门的教材。Google公司提供的TensorFlow工具就为广大学生和工程师快速迈进深度学习人工智能应用开发领域提供了跳板，本书的作者应这种需求，结合自己在人工智能领域多年的研究成果和丰富的教学经验，以深度学习的热门套件TensorFlow为基础精心撰写了这本人工智能深度学习应用开发的教材。

全书分为12章，内容包括：环境安装、TensorFlow 2.0介绍、回归问题、二分类问题、多分类问题、神经网络训练技巧、TensorFlow 2.0高级技巧、TensorBoard高级技巧、卷积神经网络经典架构、迁移学习、变分自编码器和生成式对抗网络。各章中除了讲述深度学习的理论知识与应用技术，还安排了实验章节，通过范例程序的学习让读者在实际使用TensorFlow工具的过程中快速领悟深度学习的流程：数据集的准备和预处理，建立神经网络模型，设置训练使用的优化器、损失函数和评价指标函数，设置回调函数，训练网络模型，以及验证和测试网络模型训练的结果。

本书是一本综合讲述深度学习和TensorFlow工具应用开发的教材，为了便于高校的教学或者读者自学，作者在描述深度学习原理时文字清晰且严谨，并辅以简洁明了的插图，同时提供了步骤细致的范例程序教学，让读者可以更容易理解与吸收所学的知识和技术。本书的范例程序实例都包含完整的源代码，读者可以参照这些范例程序直接上机实践和体会。

资深架构师　赵军

2020年6月

前　言

英国数学家人工智能之父Alan Turing 1950年在Oxford University Press的*Mind*期刊发表的*Computing Machinery and Intelligence*论文中提出了机器思维的概念，从此人工智能就一直是计算机科学中非常重要的领域。发展至今有70年的历史了，其中历经了多次大起大落，尤其是其中的两次"AI寒冬"，让研究者与投资人对人工智能产生了很大的疑虑。由英国伦敦Google DeepMind开发的人工智能围棋软件AlphaGo在2016年打败了顶尖职业棋手李世石，在2017年打败了排名世界第一的柯洁，之后人工智能再度成为计算机科学的"显学"，也促成了第三波的"AI崛起"。

近几年，以深度学习为核心技术的人工智能在图像、声音、决策分析等领域已超越了人类的能力。随着深度学习人工智能技术的日益成熟，许多世界级科技公司投入了大量人力和资源在深度学习人工智能上，如Google、Facebook、Uber、Airbnb、Tesla、微软、腾讯、百度、华为、滴滴、通用等，许多中小企业也投入了相当的人力资源。根据研究机构Tractica预估，全球AI市场的规模将从2018年的81亿美元成长至2025年的1 058亿美元，并且能够应用到更多产业，例如汽车、零售、医疗、商业、电信、消费、广告、法律、保险等。

作者和所领导的多媒体系统实验室团队已经在人工智能领域研究了十多年，其中包含智能汽车系统、智能居家照护系统、智能视频监控系统、智能派遣撮合系统、人工智能小秘书系统、自动驾驶计算机视觉系统，也与大学和研究机构合作开发了自动辅助驾驶次系统，项目的名称为"自动驾驶的深度学习智能感知与情境理解系统技术"，将丰富的学界研发能量导入产业界，提升自动驾驶汽车产业的研发技术。作者在IEEE和ACM顶级国际期刊上发表了三十多篇计算机视觉人工智能相关的论文。例如，在2018年和美国华盛顿大学黄正能教授合作研发了全球第一个有效的除雪与能见度增强技术，DesnowNet深度学习网络架构，刊登在图像处理领域顶尖的期刊*IEEE Transactions on Image Processing*（Impact factor: 3.735，Rank: 14/255=5.49%）上；在2019年指导越南博士生Trung-Hieu Le研发了应用在智能居家护理系统（Intelligent Homecare Systems）上的高精确度手部侦测识别技术，发表在传感器领域顶尖的期刊*IEEE Sensors Journal*（Impact factor: 2.617，Rank: 8/116=6.8%）上。

作者深刻体会到目前人工智能领域的重要性与发展性，因此精心撰写了本书，期待可以启发更多学生、工程师与研究人员快速进入深度学习人工智能领域。本书使用目前热门的深度学习套件TensorFlow，带领读者深入理解深度学习的知识与技术，并且精心设计了实践的程序教学，通过每个步骤细致的项目教学让读者可以更容易理解与吸收所学的知识和技术。从新的TensorFlow 2.0入门开始，通过12章内容，理论学习和实践应用相结合。本书撰写时使用的是TensorFlow 2.0正式版。

本书范例程序源代码可通过扫描下面的二维码获得。

如果下载有问题，可通过电子邮件联系booksaga@126.com，邮件主题为"轻松学会TensorFlow 2.0人工智能深度学习应用开发范例程序源代码"。

每年有数以万篇的深度学习研究论文发表，从中可以了解人工智能领域的日新月异与广泛应用，作者以本书介绍深度学习较重要的入门内容，让读者有能力进一步钻研更高深的深度学习知识与技术。本书中的内容如有疏漏与错误，可发送邮件给予指正与鼓励。

黄士嘉

台北科技大学电子工程系 教授

加拿大安大略理工大学 国际客座教授

IEEE Sensors Journal 国际期刊编辑

IEEE BigData Congress 国际会议主席

IEEE CloudCom Conference 国际会议主席

目　　录

第1章　环境安装 1

1.1　Python 安装 1
1.1.1　Windows 安装方法 1
1.1.2　Ubuntu 安装方法 2
1.2　TensorFlow 安装 2
1.2.1　Windows 安装方法 2
1.2.2　Ubuntu 安装方法 5
1.3　Python 扩充套件安装 8
1.4　Jupyter Notebook 9
1.4.1　Windows 安装方法 9
1.4.2　Ubuntu 安装方法 10
1.4.3　设置并建立项目 10
1.4.4　常用快捷键 11
1.4.5　Jupyter Notebook 操作练习 11
1.5　本书的范例程序 13
1.5.1　在 Windows 中打开项目 14
1.5.2　在 Ubuntu 中打开项目 14

第2章　TensorFlow 2.0 介绍 16

2.1　什么是深度学习 16
2.2　建立项目 17
2.3　TensorFlow 介绍 18
2.4　TensorFlow 2.0 的变化 20

2.5 Eager Execution ... 21
2.5.1 Eager Execution 介绍 ... 21
2.5.2 TensorFlow 基本运算 ... 22
2.6 Keras ... 24
2.6.1 Keras 介绍 ... 24
2.6.2 序贯模型 ... 26
2.6.3 Functional API ... 28
2.7 tf.data ... 32
2.7.1 tf.data 介绍 ... 32
2.7.2 基本操作 ... 34

第 3 章 回归问题 ... 39
3.1 深度神经网络 ... 39
3.1.1 神经网络简史 ... 39
3.1.2 神经网络原理 ... 40
3.1.3 全连接 ... 41
3.1.4 损失函数 MSE 和 MAE ... 41
3.1.5 神经网络权重更新 ... 43
3.1.6 神经网络训练步骤 ... 44
3.2 Kaggle 介绍 ... 46
3.3 实验一：房价预测模型 ... 47
3.3.1 数据集介绍 ... 47
3.3.2 新建项目 ... 48
3.3.3 程序代码 ... 49
3.4 TensorBoard 介绍 ... 56
3.5 实验二：过拟合问题 ... 58
3.5.1 过拟合说明 ... 58
3.5.2 程序代码 ... 60
3.5.3 TensorBoard 数据分析 ... 64
3.6 参考文献 ... 65

第 4 章　二分类问题 67

4.1　机器学习的四大类别 67
4.2　二分类问题 69
4.2.1　逻辑回归 69
4.2.2　Sigmoid 69
4.2.3　二分类交叉熵 69
4.2.4　独热编码 71
4.3　实验：精灵宝可梦对战预测 72
4.3.1　数据集介绍 72
4.3.2　新建项目 75
4.3.3　程序代码 76
4.4　参考文献 91

第 5 章　多分类问题 94

5.1　卷积神经网络 94
5.1.1　卷积神经网络简介 94
5.1.2　卷积神经网络架构 95
5.1.3　卷积神经网络的原理 102
5.2　多分类问题 105
5.2.1　Softmax 105
5.2.2　多分类交叉熵 106
5.2.3　数据增强 107
5.3　实验：CIFAR-10 图像识别 108
5.3.1　数据集介绍 108
5.3.2　TensorFlow Datasets 109
5.3.3　新建项目 110
5.3.4　程序代码 111
5.4　参考文献 127

第 6 章　神经网络训练技巧 129

6.1　反向传播 129
6.2　权重初始化 133
6.2.1　正态分布 133
6.2.2　Xavier/Glorot 初始化 135
6.2.3　He 初始化 137
6.3　批量归一化 139
6.3.1　批量归一化介绍 139
6.3.2　批量归一化网络架构 140
6.4　实验一：使用 CIFAR-10 数据集实验 3 种权重初始化方法 141
6.4.1　新建项目 141
6.4.2　建立图像增强函数 142
6.4.3　程序代码 144
6.4.4　TensorBoard 可视化权重分布 148
6.5　实验二：使用 CIFAR-10 数据集实验批量归一化方法 151
6.6　总结各种网络架构的性能比较 154
6.7　参考文献 155

第 7 章　TensorFlow 2.0 高级技巧 157

7.1　TensorFlow 高级技巧 157
7.1.1　自定义网络层 158
7.1.2　自定义损失函数 159
7.1.3　自定义评价指标函数 159
7.1.4　自定义回调函数 160
7.2　Keras 高级 API 与自定义 API 比较 161
7.2.1　网络层 161
7.2.2　损失函数 162
7.2.3　评价指标函数 163
7.2.4　回调函数 165
7.3　实验：比较 Keras 高级 API 和自定义 API 两种网络训练的结果 166

7.3.1 新建项目 166
7.3.2 程序代码 167

第 8 章 TensorBoard 高级技巧 176

8.1 TensorBoard 的高级技巧 176
8.1.1 tf.summary 177
8.1.2 tf.summary.scalar 177
8.1.3 tf.summary.image 179
8.1.4 tf.summary.text 181
8.1.5 tf.summary.audio 182
8.1.6 tf.summary.histogram 183

8.2 实验一：使用 tf.summary.image 记录训练结果 186
8.2.1 新建项目 186
8.2.2 程序代码 187

8.3 实验二：使用 TensorBoard 超参数调校工具来训练多个网络模型 195
8.3.1 启动 TensorBoard（命令行） 196
8.3.2 程序代码 197

第 9 章 卷积神经网络经典架构 205

9.1 神经网络架构 205
9.1.1 LeNet 205
9.1.2 AlexNet 206
9.1.3 VGG 206
9.1.4 GoogLeNet 207
9.1.5 ResNet 210
9.1.6 总结各种网络架构的比较 211

9.2 实验：实现 Inception V3 网络架构 212
9.2.1 新建项目 213
9.2.2 Keras Applications 214
9.2.3 TensorFlow Hub 217

9.3 参考文献 ... 222

第 10 章 迁移学习 ... 224

10.1 迁移学习 ... 224
- 10.1.1 迁移学习介绍 ... 224
- 10.1.2 迁移学习训练技巧 ... 225

10.2 实验：迁移学习范例 ... 230
- 10.2.1 新建项目 ... 230
- 10.2.2 数据集介绍 ... 231
- 10.2.3 程序代码 ... 231

10.3 参考文献 ... 237

第 11 章 变分自编码器 ... 239

11.1 自编码器介绍 ... 239
11.2 变分自编码器介绍 ... 241
11.3 变分自解码器的损失函数 ... 243
11.4 实验：变分自编码器程序代码的实现 ... 244
- 11.4.1 建立项目 ... 245
- 11.4.2 数据集介绍 ... 247
- 11.4.3 变分自编码器项目说明 ... 248
- 11.4.4 变分自编码器训练和生成图像 ... 254

11.5 参考文献 ... 257

第 12 章 生成式对抗网络 ... 258

12.1 生成式对抗网络 ... 258
- 12.1.1 生成式对抗网络介绍 ... 258
- 12.1.2 生成式对抗网络训练及损失函数 ... 260

12.2 GAN、WGAN、WGAN-GP 的演进 ... 262
- 12.2.1 生成式对抗网络的问题 ... 262
- 12.2.2 Wasserstein 距离介绍 ... 264
- 12.2.3 WGAN-GP 损失函数 ... 266

12.3 实验：WGAN-GP 程序代码的实现	268
12.3.1 建立项目	269
12.3.2 数据集介绍	271
12.3.3 WGAN-GP 项目说明	272
12.4 参考文献	281

第 1 章

环境安装

学习目标

- 在 Windows 或 Ubuntu 操作系统上安装 Python、TensorFlow 和额外的常用扩充套件
- 在 Windows 或 Ubuntu 操作系统上安装 Jupyter Notebook,并学会基本操作指令
- 下载本书所有的范例程序,并通过 Jupyter Notebook 浏览及执行程序代码

1.1 Python 安装

1.1.1 Windows 安装方法

首先,到 Python 官方网站(https://www.python.org/downloads/windows/)下载 Python 安装文件。TensorFlow 的版本要求有 Python 3.4、Python 3.5 或 Python 3.6 的支持,所以务必安装这 3 种版本的任意一种。下面将以 Python 3.6.8 为例进行介绍。

Step 01 下载 Python 3.6.8 安装文件,如图 1-1 所示。

图 1-1 下载 Python 安装文件

Step 02 安装 Python。

将 Python 加入目录路径后，单击 Install Now 按钮进行安装，如图 1-2 所示。

图 1-2　安装 Python

1.1.2　Ubuntu 安装方法

由于 Ubuntu 操作系统已经内建 Python 了，因此省略 Python 的安装步骤。

1.2　TensorFlow 安装

TensorFlow Python 的深度学习套件有许多种安装方法，例如本地安装、Virtualenv 虚拟机安装、Docker 安装等。本书使用 TensorFlow 推荐的 Virtualenv 虚拟机安装方式，并介绍两种操作系统（Windows 和 Ubuntu）和两种版本（CPU 和 GPU）的安装方式。

1.2.1　Windows 安装方法

Step 01 启动"命令提示符"程序，如图 1-3 所示。

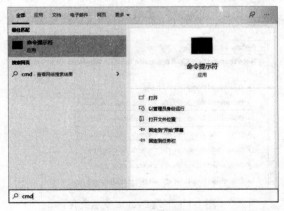

图 1-3　启动"命令提示符"程序

Step 02 安装 Virtualenv 虚拟机。

```
pip install virtualenv
```

Step 03 建立新的虚拟环境。

在下面的指令中，tf2 代表虚拟环境的名称，如果接下来要把 TensorFlow 安装为 GPU 版，那么可以将虚拟环境的名称命名为 **tf2-gpu** 以进行区别。

```
virtualenv --system-site-packages -p python ./tf2
```

Step 04 进入虚拟环境，如图 1-4 所示。

```
cd tf2\Scripts activate
```

图 1-4　进入虚拟环境

Step 05 升级 pip 版本。

```
pip install --upgrade pip
```

Step 06 安装 TensorFlow。

安装 TensorFlow 时，有两种版本可以选择，分别为 CPU 版和 GPU 版，如果安装的目标系统中有 NVIDIA GPU 并支持 CUDA，那么推荐安装 GPU 版。

- CPU 版

```
pip install --upgrade tensorflow==2.0.0-beta1
```

- GPU 版

先到 https://developer.nvidia.com/cuda-gpus 查看目标系统中的显卡是否支持 CUDA，再依次安装显卡驱动程序、CUDA10 和 cuDNN，最后安装 TensorFlow-GPU 版。

（1）安装显卡驱动程序：到 https://www.nvidia.com/drivers 下载相对应的显卡驱动程序并进行安装，如图 1-5 所示。

图 1-5　下载显卡驱动程序

（2）安装 CUDA。到 https://developer.nvidia.com/cuda-downloads 单击右下角的 Legacy Releases 按钮，如图 1-6 所示。

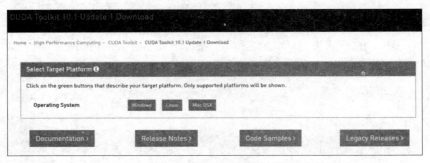

图 1-6　CUDA 下载页面

（3）选择 CUDA Toolkit 10.0 版本，如图 1-7 所示。

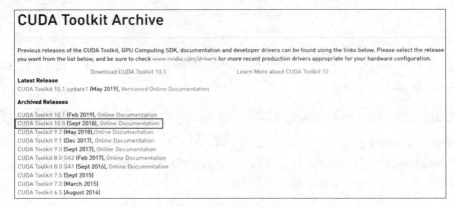

图 1-7　选择 CUDA 安装版本

（4）选择 Windows 10 的安装包，下载并进行安装，如图 1-8 所示。

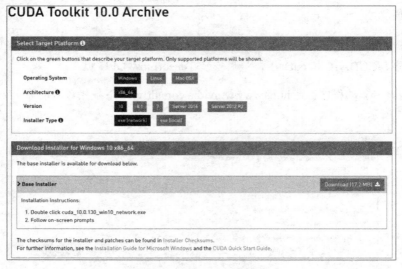

图 1-8　下载 CUDA 安装包

(5)安装 cuDNN。到 https://developer.nvidia.com/rdp/cudnn-download 下载 cuDNN for CUDA 10.0 Windows 10 版本，并进行解压缩，如图 1-9 所示。

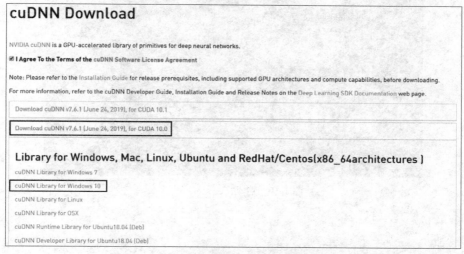

图 1-9 下载 cuDNN 安装包

(6)接下来，从 cuDNN 的解压缩目录复制 3 个文件到 CUDA 的安装目录，步骤如下：

① 把<cuDNN 解压缩目录>下的文件\cuda\bin\cudnn64_7.dll 复制到 C:\Program Files\NVIDIA GPU Computing Toolkit\CUDA\v10.0\bin 目录下。

② 把<cuDNN 解压缩目录>下的文件\cuda\include\cudnn.h 复制到 C:\Program Files\NVIDIA GPU Computing Toolkit\CUDA\v10.0\include 目录下。

③ 把<cuDNN 解压缩目录>下的文件\cuda\lib\x64\cudnn.lib 复制到 C:\Program Files\NVIDIA GPU Computing Toolkit\CUDA\v10.0\lib\x64 目录下。

(7)安装 TensorFlow GPU 版的 pip 软件包：

```
pip install --upgrade tensorflow-gpu==2.0.0-beta1
```

Step 07 验证安装。

```
python -c "import tensorflow as tf; print(tf.constant([[1, 2], [3, 4]]))"
```

1.2.2 Ubuntu 安装方法

Step 01 启动 Terminal 程序（终端程序），如图 1-10 所示。

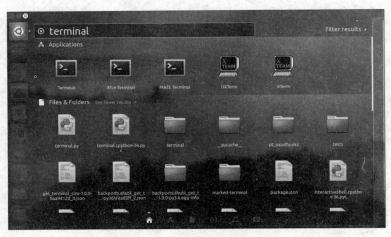

图 1-10　启动 Terminal 程序

Step 02 安装 pip。

```
sudo apt-get install python3-pip
```

Step 03 安装 Virtualenv 虚拟机。

```
sudo apt install virtualenv
```

Step 04 建立新的虚拟环境。

在指令中，tf2 代表虚拟环境的名称，如果接下来要把 TensorFlow 安装为 GPU 版，那么可以将虚拟环境的名称命名为 tf2-gpu 来进行区别。

```
virtualenv --system-site-packages -p python3 ./tf2
```

Step 05 进入虚拟环境。

```
source tf2/bin/activate
```

Step 06 升级 pip 版本。

```
pip install --upgrade pip
```

Step 07 安装 TensorFlow。

- CPU 版

```
pip install tensorflow
```

- GPU 版

NVIDIA GPU 支持检查： 先到 https://developer.nvidia.com/cuda-gpus 查看显卡是否支持 CUDA，再按照以下步骤进行安装：

（1）Ubuntu 18.04 安装方法：如果操作系统为 Ubuntu 18.04，那么到此网址 https://www.tensorflow.org/install/gpu#ubuntu_1804_cuda_10 复制下面的指令进行安装。

```
# Add NVIDIA package repositories
wget https://developer.download.nvidia.com/compute/cuda/repos/ubuntu1804/
x86_64/cuda-repo-ubuntu1804_10.0.130-1_amd64.deb
sudo dpkg -i cuda-repo-ubuntu1804_10.0.130-1_amd64.deb
sudo apt-key adv --fetch-keys https://developer.download.nvidia.com/compute/
cuda/repos/ubuntu1804/x86_64/7fa2af80.pub
sudo apt-get update
wget http://developer.download.nvidia.com/compute/machine-learning/repos/
ubuntu1804/x86_64/nvidia-machine-learning-repo-ubuntu1804_1.0.0-1_amd64.deb
sudo apt install ./nvidia-machine-learning-repo-ubuntu1804_1.0.0-1_amd64.deb
sudo apt-get update

# Install NVIDIA driver
sudo apt-get install --no-install-recommends nvidia-driver-418
# Reboot. Check that GPUs are visible using the command: nvidia-smi

# Install development and runtime libraries (~4GB)
sudo apt-get install --no-install-recommends \
    cuda-10-0 \
    libcudnn7=7.6.0.64-1+cuda10.0 \
    libcudnn7-dev=7.6.0.64-1+cuda10.0

# Install TensorRT. Requires that libcudnn7 is installed above.
sudo apt-get update && \
    && sudo apt-get install -y --no-install-recommends libnvinfer-dev=5.1.5-1
+cuda10.0
```

（2）Ubuntu 16.04 安装方法：如果操作系统为 Ubuntu 16.04，那么到此网址 https://www.tensorflow.org/install/gpu#ubuntu_1604_cuda_10 复制下面的指令进行安装。

```
# Add NVIDIA package repositories
# Add HTTPS support for apt-key
sudo apt-get install gnupg-curl
wget https://developer.download.nvidia.com/compute/cuda/repos/ubuntu1604/
x86_64/cuda-repo-ubuntu1604_10.0.130-1_amd64.deb
sudo dpkg -i cuda-repo-ubuntu1604_10.0.130-1_amd64.deb
sudo apt-key adv --fetch-keys https://developer.download.nvidia.com/compute/
cuda/repos/ubuntu1604/x86_64/7fa2af80.pub
sudo apt-get update
wget http://developer.download.nvidia.com/compute/machine-learning/repos/
ubuntu1604/x86_64/nvidia-machine-learning-repo-ubuntu1604_1.0.0-1_amd64.deb
sudo apt install ./nvidia-machine-learning-repo-ubuntu1604_1.0.0-1_amd64.deb
sudo apt-get update

# Install NVIDIA driver
# Issue with driver install requires creating /usr/lib/nvidia
sudo mkdir /usr/lib/nvidia
sudo apt-get install --no-install-recommends nvidia-410
# Reboot. Check that GPUs are visible using the command: nvidia-smi

# Install development and runtime libraries (~4GB)
```

```
sudo apt-get install --no-install-recommends \
    cuda-10-0 \
    libcudnn7=7.4.1.5-1+cuda10.0 \
    libcudnn7-dev=7.4.1.5-1+cuda10.0

# Install TensorRT. Requires that libcudnn7 is installed above.
sudo apt-get update && \
    sudo apt-get install nvinfer-runtime-trt-repo-ubuntu1604-5.0.2-ga-cuda10.0 \
    && sudo apt-get update \
    && sudo apt-get install -y --no-install-recommends libnvinfer-dev=5.0.2-1+cuda10.0
```

（3）安装 TensorFlow GPU 版的 pip 软件包。

```
pip install tensorflow-gpu
```

Step 08 验证安装。

```
python -c "import tensorflow as tf; print(tf.constant([[1, 2], [3, 4]]))"
```

> **说 明**
>
> 与直接安装在 Python 本地端的环境中相比，使用 Virtualenv 虚拟机或 Docker 的安装方法能够建立多个环境，并在环境中安装不同版本的 TensorFlow，环境之间不会相互影响，所以推荐读者使用 Virtualenv 虚拟机或 Docker 的安装方法。

1.3　Python 扩充套件安装

Windows 与 Ubuntu 的安装方法相同。首先，启动 Terminal（Ubuntu）程序或"命令提示符"（Windows）程序，并进入 Virtualenv 虚拟环境中，在窗口中输入下面的指令。

- NumPy

数学函数库，支持矩阵运算和大量数学函数，也支持许多相关工具，例如 Matplotlib 或 OpenCV 都支持 NumPy 数据类型，比 TensorFlow 的数据类型更为方便。

```
pip install numpy
```

- Matplotlib

数据可视化工具，可以用来绘制各种图表，例如折线图、柱状图和三维图等。

```
pip install matplotlib
```

- Pandas

数据分析以及可视化工具，提供数据处理和数据可视化工具，例如读取 Excel 文件并对其进行统计、运算或可视化。

```
pip install pandas
```

- OpenCV

计算机视觉图像处理工具，提供图像读取、绘制、摄影机图像读取或图像处理算法等工具。

```
pip install opencv-python
```

- TensorFlow Datasets

拥有多个开放数据集的数据库，便于使用者下载和使用数据集。

```
pip install tensorflow-datasets
```

- TensorFlow Hub

拥有多个开放网络模型和预训练权重，便于使用者搭建网络模型和使用预训练权重。

```
pip install tensorflow-hub
```

- TensorFlow Addons

为 TensorFlow 的扩充套件，里面有许多 TensorFlow 不具备的功能。这些新功能可能是因为广泛使用性不足、只有少部分领域会使用或新发展的技术等原因，所以并没有被加入 TensorFlow 中。

```
pip install tensorflow-addons
```

> **说 明**
>
> 目前 TensorFlow Addons 只支持 Ubuntu 系统，不支持 Windows 系统。详细信息可参考 https://github.com/tensorflow/addons/issues/173。

1.4 Jupyter Notebook

Jupyter Notebook 是互动性和可视化的集成开发环境（IDE），每执行一段程序就会返回输出，非常适合初学者使用。Jupyter Notebook 主要支持 3 种程序设计语言 Julia、Python 和 R，这也是它名字的由来，不过到目前为止可支持的程序设计语言已经超过 40 多种。本节介绍 Jupyter Notebook 在两种操作系统（Windows 和 Ubuntu）中的安装及使用方法。

1.4.1 Windows 安装方法

Step 01 安装 Jupyter Notebook。

```
python -m pip install jupyter
```

Step 02 将虚拟环境加入 Jupyter Notebook 中。

在指令中，tf2 代表虚拟环境名称。

```
.\tf2\Scripts\activate
pip install ipykernel
python -m ipykernel install --name=tf2
```

1.4.2 Ubuntu 安装方法

Step 01 安装 Jupyter Notebook。

```
python3 -m pip install jupyter
```

Step 02 将虚拟环境加入 Jupyter Notebook 中。

在指令中，tf2 代表虚拟环境名称。

```
source tf2/bin/activate
pip3 install ipykernel
python3 -m ipykernel install --user --name=tf2
```

1.4.3 设置并建立项目

Step 01 启动 Jupyter Notebook。

```
jupyter notebook
```

Step 02 新建执行文件。

单击界面右上角的 New 下拉菜单，然后单击刚刚安装的 Python 解释器（在 Jupyter 中都称为 Kernel）来启动它，图 1-11 显示了 3 个不同的 Kernel（内核），分别为：

- Python3：本地端 Python。
- tf2：虚拟机 Python（安装 TensorFlow-cpu 版本）。
- tf2-gpu：虚拟机 Python（安装 TensorFlow-gpu 版本）。

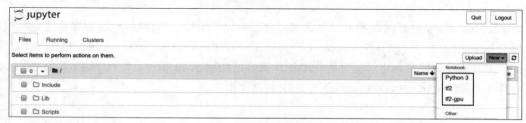

图 1-11 新建执行文件

Step 03 选择 Kernel。

进入 Python 文件后，可以依次单击菜单选项 Kernel→Change kernel 来选择 Kernel，界面右上角会显示当前使用的是哪个 Kernel，如图 1-12 所示。

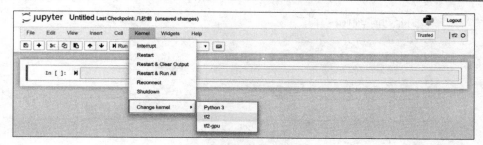

图 1-12　选择 Kernel

1.4.4　常用快捷键

常用快捷键如表 1-1 所示。

表 1-1　常用快捷键

快 捷 键	说　明
A	增加一行
D + D	减少一行
M	转为 Markdown
Y	转为 Code
Shift + Enter	执行单行指令
Shift + Tab	查询函数参数
Esc	退出编辑模式
Enter	进入编辑模式

1.4.5　Jupyter Notebook 操作练习

Jupyter Notebook 有两种模式：编辑模式（Edit Mode）和命令模式（Command Mode），在不同模式下可以进行不同的操作。

Step 01　图 1-13 所示为编辑模式的状态。编辑模式为编写程序代码或文字的模式，可以使用 Shift + Enter 或 Shift + Tab 等快捷键。

图 1-13　编辑模式

Step 02　图 1-14 所示为命令模式的状态。大多数快捷键都只能在命令模式下使用,例如 A 、 D + D 、 M 、 Y 或 Shift + Enter 等。

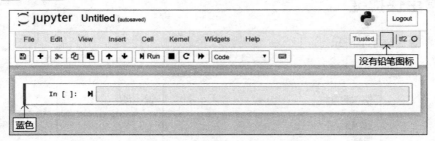

图 1-14　命令模式

Step 03　先在编辑模式下输入 print("Hello Jupyter NoteBook"),再按 Shift + Enter 快捷键来执行程序代码,如图 1-15 所示。

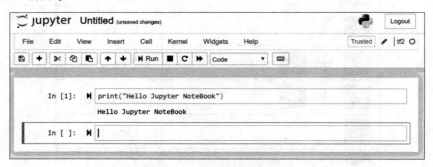

图 1-15　命令模式执行程序代码

Step 04　按 Esc 快捷键跳出编辑模式,进入命令模式,如图 1-16 所示。

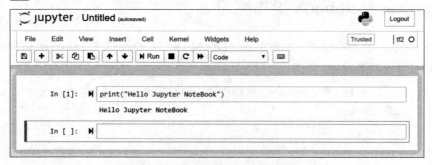

图 1-16　使用快捷键退出编辑模式

$\begin{smallmatrix}\text{Step}\\\text{05}\end{smallmatrix}$ 自行尝试其他指令，熟悉一下编辑环境，例如 A 、 D + D 、 M 或 Shift + Enter 等，更多快捷键可以通过图 1-17 所示的键盘图标进行查询和设置。

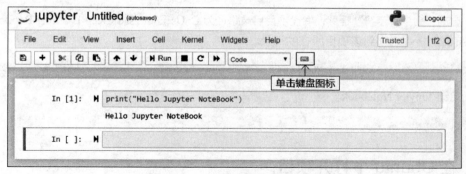

图 1-17 快捷键查询和设置

1.5 本书的范例程序

本书的范例程序可扫描前言的二维码下载。下载完成后，将压缩文件解压缩，即可得到如图 1-18 所示的范例程序相关文件。

名称	修改日期	类型	大小
audio	2020/11/11 10:59	文件夹	
image	2020/11/11 10:59	文件夹	
.gitignore	2020/11/11 10:59	GITIGNORE 文件	1 KB
Cover	2020/11/11 10:59	JPG 文件	83 KB
debug	2020/11/11 11:20	文本文档	1 KB
Lab1.ipynb	2020/11/11 18:36	IPYNB 文件	114 KB
Lab2.ipynb	2020/11/11 18:11	IPYNB 文件	337 KB
Lab3.ipynb	2020/11/11 17:53	IPYNB 文件	1,214 KB
Lab4.ipynb	2020/11/11 17:53	IPYNB 文件	1,062 KB
Lab5.ipynb	2020/11/11 18:40	IPYNB 文件	199 KB
Lab6.ipynb	2020/11/12 1:40	IPYNB 文件	64 KB
Lab7.ipynb	2020/11/12 1:35	IPYNB 文件	630 KB
Lab8.ipynb	2020/11/12 1:09	IPYNB 文件	345 KB
Lab9.ipynb	2020/11/11 18:52	IPYNB 文件	138 KB
LICENSE	2020/11/11 10:59	文件	2 KB
preprocessing	2020/11/11 12:27	Python File	3 KB
test	2020/11/11 10:59	Python File	0 KB

图 1-18 本书的范例程序

1.5.1 在 Windows 中打开项目

Step 01 启动"命令提示符"程序,假设项目目录路径为 C:\Users\mmslab\Deep-Learning-Book-master,需要通过下面的指令切换到该项目所在的目录。

```
cd C:\Users\mmslab\Deep-Learning-Book-master
```

Step 02 启动 Jupyter Notebook。

```
jupyter notebook
```

1.5.2 在 Ubuntu 中打开项目

Step 01 启动 Terminal 程序,假设项目目录路径为/home/mmslab/Deep-Learning-Book-master,需要通过下面的指令切换到该项目所在的目录。

```
cd /home/mmslab/Deep-Learning-Book-master
```

Step 02 启动 Jupyter Notebook,结果如图 1-19 所示。

```
jupyter notebook
```

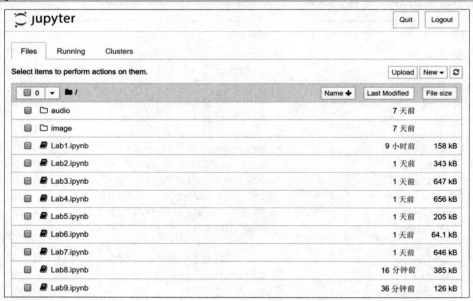

图 1-19 通过 Jupyter Notebook 打开项目

Step 03 单击要打开的章节的文件,例如要打开第 1 章的程序代码,就单击 Lab1.ipynb 文件,如图 1-20 所示。

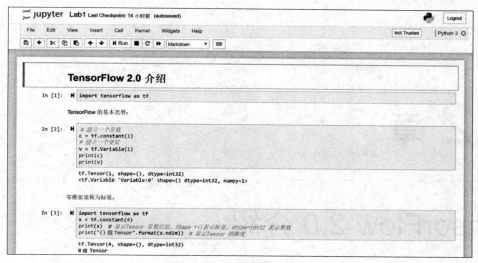

图 1-20　第 1 章程序的页面

Step 04　通过快进图标执行整个程序项目，如图 1-21 所示。

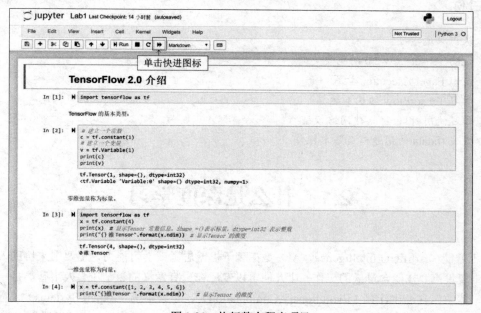

图 1-21　执行整个程序项目

第 2 章

TensorFlow 2.0 介绍

学习目标

- 认识深度学习
- 说明 TensorFlow 2.0 的重要更新
- 认识 Eager Execution 模式
- 学习 TensorFlow 的基本运算
- 学习使用 tf.keras 搭建网络模型
- 概述 tf.data 的用途以及基本操作

2.1 什么是深度学习

人工智能（Artificial Intelligence，AI）是指赋予机器思考或学习能力，主要用来协助人类或取代那些重复性高、技能含量低的工作，让人们可以专注在更有意义的事情上。人工智能是一个综合的领域，其中包含进化计算（Evolutionary Computation）、专家系统（Expert Systems）、符号人工智能（Symbolic Artificial Intelligence）、支持向量机（Support Vector Machine）、机器学习（Machine Learning）、深度学习（Deep Learning）、强化学习（Reinforcement Learning）等众多领域，而深度学习又是机器学习的一个分支领域，如图 2-1 所示。

在深度学习中，可以将深度学习网络视为一个复杂函数，当输入一组训练数据到神经网络时，可得到一组输出预测，接着根据输出预测与标记答案的差距更新神经网络的权重（Weight），使其更加逼近于期望的输出，即标记答案。下面我们将介绍热门和流行的人工智能工具 TensorFlow。

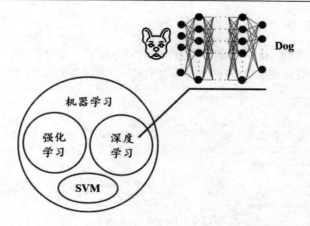

图 2-1　深度学习是目前机器学习热门的领域

2.2　建立项目

建议使用 Jupyter Notebook 来执行本章的程序代码，操作流程如下：

Step 01 启动 Jupyter Notebook。

在 Terminal（Ubuntu）或命令提示符（Windows）中输入以下指令：

```
.\tf2\Scripts\activate
jupyter notebook
```

Step 02 新建执行文件。

单击界面右上角的 New 下拉按钮，然后单击所安装的 Python 解释器（在 Jupyter 中都称为 Kernel）来启动，如图 2-2 所示，显示了 3 个不同的 Kernel，分别为：

- Python 3：本地端 Python。
- tf2：虚拟机 Python（TensorFlow-cpu 版本）。
- tf2-gpu：虚拟机 Python（TensorFlow-gpu 版本）。

图 2-2　新建执行文件

Step 03 执行程序代码。

在方框中输入程序代码"print("Hello Jupyter Notebook")",再按 Shift + Enter 快捷键来执行单行程序代码,执行结果会显示在该行程序代码的下方,如图 2-3 所示。

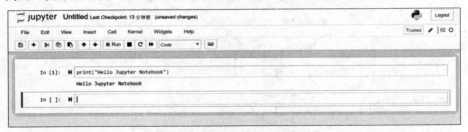

图 2-3　Jupyter 环境界面

Step 04 载入 TensorFlow 套件。

在方框中输入"import tensorflow as tf",再按 Shift + Enter 快捷键来执行单行程序代码,即可载入 TensorFlow 套件,如图 2-4 所示。

图 2-4　载入 TensorFlow 套件

接下来,本章的范例程序代码都可在 Jupyter Notebook 上执行。

2.3　TensorFlow 介绍

DistBelief 为 Google Brain 团队一开始所使用的机器学习工具,后来以 DistBelief 为基础做了改进,并开放了源代码,才有了现在我们所熟知的 TensorFlow。目前,TensorFlow 已成为流行的机器学习工具之一,而 TensorFlow 的命名也直接反映了其功能,可以将名字拆成 Tensor 和 Flow 两部分来解读。

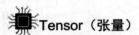Tensor(张量)

张量是矩阵向任意维度(Dimension)的推广,TensorFlow 的运算都是基于张量进行的。

- TensorFlow 的基本类型。

```
import tensorflow as tf
```

```
# 建立一个常数 Tensor
c = tf.constant(1)
# 建立一个变量 Tensor
v = tf.Variable(1)
print(c)   # 显示 Tensor 常数信息，Shape=() 表示标量，dtype=int32 表示整数
print(v)   # 显示 Tensor 变量信息，Shape=() 表示标量，dtype=int32 表示整数
```

结果如下：

```
tf.Tensor(1, shape=(), dtype=int32)
<tf.Variable 'Variable:0' shape=() dtype=int32, numpy=1>
```

- 零维张量称为"标量"。

```
x = tf.constant(4)
print(x)   # 显示 Tensor 常数信息，Shape=() 表示标量，dtype=int32 表示整数
print("{}维 Tensor".format(x.ndim))     # 显示 Tensor 的维度
```

结果如下：

```
tf.Tensor(4, shape=(), dtype=int32)
0 维 Tensor
```

- 一维张量称为"向量"。

```
x = tf.constant([1, 2, 3, 4, 5, 6])
print("{}维 Tensor ".format(x.ndim))     # 显示 Tensor 的维度
```

结果如下：

```
1 维 Tensor
```

- 二维张量称为"矩阵"。

```
x = tf.constant([[1, 2, 3], [4, 5, 6]])
print("{}维 Tensor ".format(x.ndim))     # 显示 Tensor 的维度
```

结果如下：

```
2 维 Tensor
```

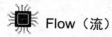 Flow（流）

流解释为数据流动或计算。

TensorFlow 的运作方式是通过产生数据流图（Data Flow Graph）来进行运算，又称作"计算图"（Computation Graph）。计算图的节点（Node）用来表示数学运算，边（Edge）则表示节点间的关联性。图 2-5 所示为数学式 ReLU(XW+b) 的计算图。

TensorFlow 1.x 以前的版本都是先产生计算图，再执行如上所述的运算，此架构被称为静态图。而后来的 TensorFlow 2.0 由于默认 Eager Execution 模式为执行模式，因此引入了动态图的机制，使得执行指令可以立即得到回复。

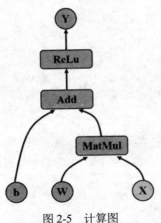

图 2-5 计算图

2.4 TensorFlow 2.0 的变化

TensorFlow 2.0 和前一版相比更简洁且容易上手。下面开始介绍在 TensorFlow 2.0 中当前主要推行的 6 个功能，其中 Eager Execution、Keras 和 tf.data 这 3 个功能将在接下来的 3 节详细介绍，而另外 3 个功能在后面的章节中再进行详细说明。

- Eager Execution

动态图模式，表示立即执行的意思，在该模式下执行运算会立即返回数值，让开发者的调试（Debug）更加便利和快速。

- Keras

TensorFlow 2.0 加入了 Keras 作为内建的高级 API 后有了更高的兼容性。内建 Keras 可通过 tf.keras 方式来调用其功能，具有指令简洁、可以自由组合且容易扩展的模块化 API 等特性，使得神经网络更容易搭建。

- tf.data

使用 tf.data 建立数据输入管道（Input Pipeline），速度更快，更简单。

- TensorFlow Hub

共享模型权重的 Library，可以从 Hub 上加载预先训练好的模型（Model），也可以将自己训练好的模型上传到 Hub，以分享给其他人。

- Distribution Strategy

新的 API 可用于更加轻松地完成在多台设备上的分布式训练，例如 CPU、GPU 或 TPU。

- SavedModel

TensorFlow 2.0 已经规范好网络模型的存储格式，使我们可以将训练好的网络模型放到想要执行的平台上，例如手机、树莓派或网页，同时也支持不同的程序设计语言，例如 C、Java、Go 或 C#等。

图 2-6 展示了当前 TensorFlow 2.0 已经能包办从训练模型到部署模型的流程，即读取数据、训练模型、保存模型以及把模型部署到各种设备的平台上。

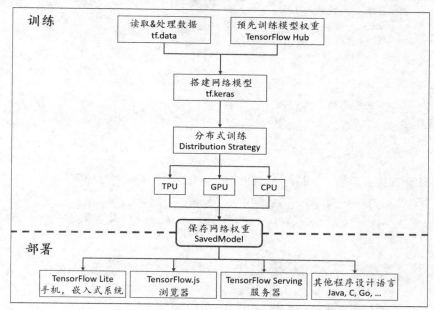

图 2-6　模型训练及部署流程所使用的 TensorFlow 模块

2.5　Eager Execution

2.5.1　Eager Execution 介绍

TensorFlow 引入了 Eager Execution 动态图模式，这个模式在 TensorFlow 2.0 中为默认的执行模式，一旦执行运算就会立刻返回数值，这有别于以往的静态图模式，需要建立计算图才能执行。如此使得 TensorFlow 更容易入门学习，也使得程序开发更为直观。

Eager Execution 模式的优点如下：

（1）立即返回数值，方便调试。
（2）不必定义计算图。
（3）不必初始化参数。
（4）无须 tf.Session.run 就可以返回运算结果。

TensorFlow 1.x 和 TensorFlow 2.0 的比较：

- TensorFlow 1.x code

tf.constant 在计算图中建立节点，并可以通过 sess.run 从中取得数值：

```
a = tf.constant(1)         # 建立一个常数 Tensor
print(a)                   # 显示 Tensor 常数信息，Shape=() 表示标量，dtype=int32 表示整数
# 建立一个 Session
sess = tf.Session()
# 通过 sess.run 取数值并显示出来
print("a = {}".format(sess.run(a)))
# 关闭 Session 释放资源
sess.close()
```

结果如下：

```
Tensor("Const_5:0", shape=(), dtype=int32)
a = 1
```

- TensorFlow 2.0 code

tf.constant 会直接返回数值，相对于 TensorFlow 1.x 省去了许多行程序代码：

```
a = tf.constant(1)         # 建立一个常数 Tensor
print(a)                   # 显示 Tensor 常数信息，Shape=() 表示标量，dtype=int32 表示整数
```

结果如下：

```
tf.Tensor(1, shape=(), dtype=int32)
```

2.5.2 TensorFlow 基本运算

Step 01 导入必要的套件。

```
import numpy as np           # 载入 NumPy 数学函数库
import tensorflow as tf      # 载入 TensorFlow 深度学习函数库
# 检查 Eager Execution 模式是否启动
print("Eager Execution 是否启动: {}".format(tf.executing_eagerly()))
```

结果如下：

```
Eager Execution 是否启动:True
```

Step 02 定义常数 Tensor。

```
a = tf.constant(3)     # 建立一个常数为 3 的 Tensor
b = tf.constant(4)     # 建立一个常数为 4 的 Tensor
# 显示 Tensor 数值(format 会直接将数值赋值给"{}"，不会有 shape、dtype 等信息)
print("a = {}".format(a))
print("b = {}".format(b))
```

结果如下：

```
a = 3
b = 4
```

Step 03 检查数据类型。

```
print(a)    # 显示 Tensor 常数信息，Shape=()表示标量，dtype=int32 表示整数
print(b)    # 显示 Tensor 常数信息，Shape=()表示标量，dtype=int32 表示整数
```

结果如下：

```
tf.Tensor(3, shape=(), dtype=int32)
tf.Tensor(4, shape=(), dtype=int32)
```

Step 04 基本运算。

```
c = a + b
print("a + b = {}".format(c))    # 显示 a+b 的结果
d = a * b
print("a * b = {}".format(d))    # 显示 a*b 的结果
```

结果如下：

```
a + b = 7
a * b = 12
```

Step 05 二维 Tensor 的运算。

在 Eager Execution 模式下，可以混合 Tensor 和 NumPy 进行运算。

```
# 建立二维张量的 Tensor，并且 dtype 为 float32
a = tf.constant([[1., 2.], [3., 4.]], dtype=tf.float32)
# 建立一个 NumPy array 数组，并且 dtype 为 float32
b = np.array([[1., 0.], [2., 3.]], dtype=np.float32)
print("a constant: {}D Tensor".format(a.ndim))

c = a + b
print("a + b = \n{}".format(c))    # 显示 a+b 的结果
# tf.matmul 为矩阵乘法
d = tf.matmul(a, b)
print("a * b = \n{}".format(d))    # 显示 a*b 的结果
```

结果如下：

```
a constant: 2D Tensor
a + b = [[2. 2.]
         [5. 7.]]
a * b = [[ 5.  6.]
         [11. 12.]]
```

Step 06 输出的结果为 Tensor 格式，可以将它转为 NumPy 格式。

```
print("NumpyArray:\n {}".format(c.numpy()))
```

结果如下：

```
NumpyArray:
 [[2. 2.]
 [5. 7.]]
```

> **说　明**
>
> 在 TensorFlow 2.0 中，虽然有 Eager Execution 模式能够让 Tensor 格式支持 Python 基本运算（ex: for、if…else 等），但并非完全兼容，例如 OpenCV 或 Matplotlib 等套件的 API 输入格式有可能会不支持 Tensor 格式。而当遇到数据类型错误等问题时，最快的解决方法是将其转为 NumPy 格式，因为其历史悠久、高性能及受欢迎程度高，让 NumPy 格式能适用于大部分的 Python 套件。

Step 07 计算梯度：假设损失函数为 w^2。

```
w = tf.Variable([[1.0]])
# 正向传播会被记录到"tape"中
with tf.GradientTape() as tape:
    loss = w * w  # 损失函数为 w²

# 反向传播"tape"计算梯度，∂loss²/∂w = ∂w²/∂w = 2w，因为 w=1，所以 grad=2

grad = tape.gradient(loss, w)
print(grad)
```

结果如下：

```
tf.Tensor([[2.]], shape=(1, 1), dtype=float32)
```

> **说　明**
>
> 简单来说，一维的标量 x 的梯度（Gradient）就是计算 f(x) 对 x 的微分，同理多维的向量 x 的梯度就是计算 f(x) 对所有元素的偏微分。计算出来的梯度是有方向和大小的向量，而梯度指向的方向为局部最大值，所以在第 3 章介绍的梯度下降法就是往梯度的反方向（局部最小值）更新权重。

2.6　Keras

2.6.1　Keras 介绍

Keras 是 François Chollet 于 2014—2015 年开始编写的开源高级深度学习 API，主要用于快速搭建和训练网络模型。Keras 本身并没有运算能力，它是在 TensorFlow、CNTK 和 Theano 等深度学习开源套件上执行的，这些套件称为 Keras 的后端（Backend）。开始时 Keras 后端只支持 Theano，直到 2015 年底 TensorFlow 开源后，Keras 才搭建了 TensorFlow 后端，而今天 TensorFlow 已成为

Keras 常用的后端。总而言之，Keras 将这些深度学习套件封装为更容易使用的指令。

TensorFlow 2.0 将 Keras 收纳为内建的高级 API，因此不用额外安装 Keras 套件，直接通过 tf.keras 指令即可调用。tf.keras 和 Keras 的差别在于，tf.keras 能更为全面地支持 TensorFlow 的指令与模式，例如支持 Eager Execution、tf.data、TPU 训练等。

下面将介绍两种常用的网络搭建方法：

（1）Sequential Model（序贯模型）。
（2）Functional API（函数式模型，或称为函数式 API）。

接下来的范例会使用以下几种网络层，先简略介绍。暂时没有弄明白也没有关系，详细的理论和功能会在后面的章节讲述。以下范例展示的重点在于使用 Keras 搭建网络架构的方便性与灵活性。

（1）Dense：搭建全连接层的指令。
（2）Conv2d：搭建卷积层的指令。
（3）Flatten：将输入压平，重塑（Reshape）成一维张量，大多用于卷积层与全连接层之间。
（4）Add：将两个层的输出加在一起。
（5）Concatenate：将两个层的输出以指定的维度进行拼接（Concat），这种方法与 Add 方法相比保留了更多信息量，但计算量较大。

> **说　明**
>
> 使用 tensorflow.keras.utils.plot_model 需要额外安装 pydot 和 graphviz 套件。
> （1）在 Ubuntu 系统中的安装方法：启动 Terminal 程序并输入以下指令：
>
> ```
> pip install pydot
> apt-get install graphviz
> ```
>
> （2）在 Windows 系统中的安装方法：
> ① 启动"命令提示符"程序并输入以下指令：
>
> ```
> pip install pydot
> ```
>
> ② 到 https://graphviz.gitlab.io/_pages/Download/Download_windows.html 下载 Windows 系统下的安装包并安装，如图 2-7 所示。

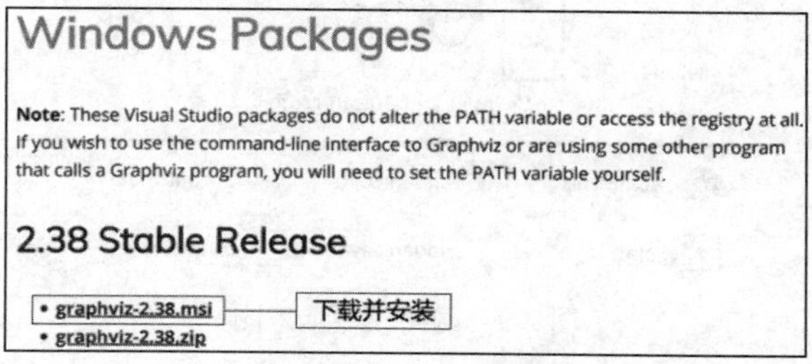

图 2-7　graphviz 下载页面

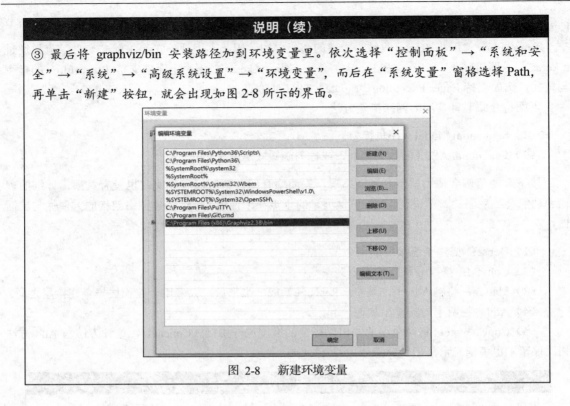

图 2-8 新建环境变量

2.6.2 序贯模型

序贯模型（Sequential Model）的搭建方法简单快速，可以解决大多数问题以及运用到各种应用中，例如手写数字识别、房价预测、评论分类等，基本上只要是回归问题或分类问题，就可以用序贯模型解决。但是，序贯模型的搭建方法有限制，必须逐层按序搭建网络，而且网络模型必须为单个输入层和单个输出层。图 2-9 所示为 Single Input and Output Model（单输入输出模型）。

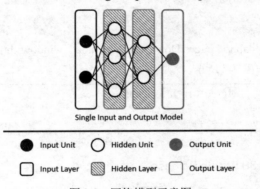

图 2-9 网络模型示意图

下面示范序贯模型的两种搭建方法，以分类问题为例：输入为 28×28 的图像并压平为 784 的一维向量，输出为 10 个元素的一维向量（分为 10 个类别），中间使用两个隐藏层，各拥有 64 个神经元，而隐藏层的激活函数为 ReLU，输出层为 Softmax。

第 2 章　TensorFlow 2.0 介绍

Step 01 导入必要的套件。

```
import tensorflow as tf
from tensorflow import keras
from tensorflow.keras import layers
from tensorflow.keras.utils import plot_model
from IPython.display import Image
```

Step 02 建立模型（搭建方法有两种）。

- 方法一

```
# 建立一个序贯模型
model = keras.Sequential(name='Sequential')
# 每次 model.add 会增加一层网络到模型中，第一层需要定义输入尺寸（input_shape）
# 第一个数值表示输出个数，即该层神经元的个数
model.add(layers.Dense(64, activation='relu', input_shape=(784,)))
model.add(layers.Dense(64, activation='relu'))
# 最后一层会被当作模型的输出层
model.add(layers.Dense(10, activation='softmax'))
```

- 方法二

```
# 可以将所有网络层放到一个列表（list）中，并作为 tf.keras.Sequential 的参数
# 而这个列表同样有顺序性，第一个需要定义输入大小，最后一个为输出层
model = tf.keras.Sequential([
                layers.Dense(64, activation='relu', input_shape=(784,)),
                layers.Dense(64, activation='relu'),
                layers.Dense(10, activation='softmax')])
```

Step 03 显示网络模型。

```
# 产生网络拓扑图
plot_model(model, to_file='Functional_API_Sequential_Model.png')

# 显示出网络拓扑图
Image('Functional_API_Sequential_Model.png')
```

结果如图 2-10 所示。

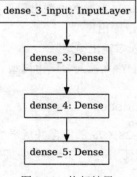

图 2-10　执行结果

2.6.3 Functional API

Keras 中的 Functional API（函数式 API）是建立网络模型的另一种方式，它提供了更多的灵活性，能够建立更复杂的模型，例如多输入单输出模型、单输入多输出模型和多输入多输出模型等，如图 2-11 所示。

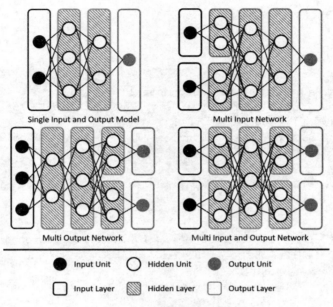

图 2-11　网络模型示意图

Functional API 在使用上非常灵活，后面介绍的高级应用都能使用 Functional API 方式搭建模型来实现。下面介绍几种高级应用。

（1）物体检测（Object Detection）。
（2）图像分割（Image Segmentation）。
（3）生成式对抗网络（Generative Adversarial Network）。

下面将按序介绍几种网络架构以及应用的场景。

（1）Single Input and Output Model：单输入单输出模型。
（2）Multi Input Model：多输入单输出模型。
（3）Multi Output Model：单输入多输出模型。
（4）Multi Input and Output Model：多输入多输出模型。

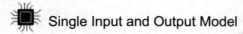　Single Input and Output Model

单输入单输出模型，例如输入为 $28×28$ 的图像并压平为 784 的一维向量，输出为 10 个元素的一维向量（分为 10 个类别），中间使用两个隐藏层，各拥有 64 个神经元，而隐藏层的激活函数为 ReLU，输出层为 Softmax。

```
# 不同于序贯模型和Functional API需要建立输入层
inputs = keras.Input(shape=(784,), name='Input')
h1 = layers.Dense(64, activation='relu', name='hidden1')(inputs)
h2 = layers.Dense(64, activation='relu', name='hidden2')(h1)
outputs = layers.Dense(10, activation='softmax', name='Output')(h2)

# 这个keras.Model会自动将输入层到输出层所有经过的各层连接起来建立成网络
model = keras.Model(inputs=inputs, outputs=outputs)

plot_model(model, to_file='Functional_API_Single_Input_And_Output_Model.png')
Image('Functional_API_Single_Input_And_Output_Model.png')
```

结果如图 2-12 所示。

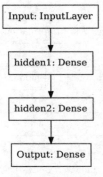

图 2-12　执行结果

Multi Input Model

多输入单输出模型，例如商品价格预测为两个输入（商品图片和商品品牌）和一个输出（价格预测），商品图片(128, 128, 3)输入经过 3 个隐藏层，商品品牌(1,)输入经过一个隐藏层，结合两个信息后再经过一个隐藏层，而输出为价格预测(1,)。

```
# 网络模型输入层
img_input = keras.Input(shape=(128, 128, 3), name='Image_Input')
info_input = keras.Input(shape=(1, ), name='Information_Input')

# 网络模型隐藏层
h1_1 = layers.Conv2D(64, 5, strides=2, activation='relu',
                     name='hidden1_1')(img_input)
h1_2 = layers.Conv2D(32, 5, strides=2, activation='relu',
                     name='hidden1_2')(h1_1)
h1_2_ft = layers.Flatten()(h1_2)
h1_3 = layers.Dense(64, activation='relu', name='hidden1_3')(info_input)
concat = layers.Concatenate()([h1_2_ft, h1_3])
h2 = layers.Dense(64, activation='relu', name='hidden2')(concat)

# 网络模型输出层
outputs = layers.Dense(1, name='Output')(h2)
```

```
# 建立网络模型
model = keras.Model(inputs=[img_input, info_input], outputs=outputs)

# 显示网络模型架构
plot_model(model, to_file='Functional_API_Multi_Input_Model.png')
Image('Functional_API_Multi_Input_Model.png')
```

结果如图 2-13 所示。

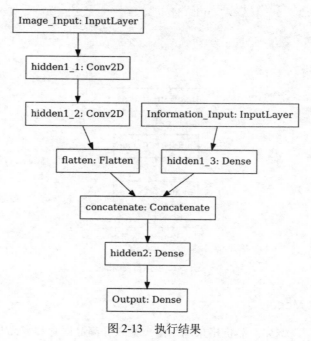

图 2-13　执行结果

Multi Output Model

单输入多输出模型,例如人像识别为一个输入(人物照片)和两个输出(年龄和性别),人物照片(128, 128, 3)输入经过 4 个隐藏层,而输出为年龄(1,)和性别(1,)两种不同的信息。

```
# 网络模型输入层
inputs = keras.Input(shape=(28, 28, 1), name='Input')

# 网络模型隐藏层
h1 = layers.Conv2D(64, 3, activation='relu', name='hidden1')(inputs)
h2 = layers.Conv2D(64, 3, strides=2, activation='relu', name='hidden2')(h1)
h3 = layers.Conv2D(64, 3, strides=2, activation='relu', name='hidden3')(h2)
flatten = layers.Flatten()(h3)

# 网络模型输出层
age_output = layers.Dense(1, name='Age_Output')(flatten)
gender_output = layers.Dense(1, name='Gender_Output')(flatten)

# 建立网络模型
```

```
model = keras.Model(inputs=inputs, outputs=[age_output, gender_output])

# 显示网络模型架构
plot_model(model, to_file='Functional_API_Multi_Output_Model.png')
Image('Functional_API_Multi_Output_Model.png')
```

结果如图 2-14 所示。

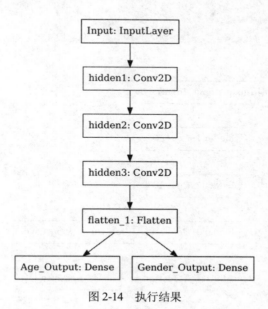

图 2-14 执行结果

Multi Input and Output Model

多输入多输出模型，例如天气预测为两个输入（卫星云图和气候信息）和 3 个输出（概率、温度和湿度），卫星云图(256×256×3)输入经过 3 个隐藏层，气候信息(10,)输入经过一个隐藏层并结合两个信息，而输出为气候信息，如降雨概率(1,)、温度(1,)和湿度(1,) 3 种不同的信息。

```
# 网络模型输入层
image_inputs = keras.Input(shape=(256, 256, 3), name='Image_Input')
info_inputs = keras.Input(shape=(10, ), name='Info_Input')

# 网络模型隐藏层(Image Input)
h1 = layers.Conv2D(64, 3, activation='relu', name='hidden1')(image_inputs)
h2 = layers.Conv2D(64, 3, strides=2, activation='relu', name='hidden2')(h1)
h3 = layers.Conv2D(64, 3, strides=2, activation='relu', name='hidden3')(h2)
flatten = layers.Flatten()(h3)
# 网络模型隐藏层(Information Input)
h4 = layers.Dense(64)(info_inputs)
concat = layers.Concatenate()([flatten, h4])  # 结合 Image 和 Information 特征

# 网络模型输出层
weather_outputs = layers.Dense(1, name='Output1')(concat)
temp_outputs = layers.Dense(1, name='Output2')(concat)
humidity_outputs = layers.Dense(1, name='Output3')(concat)
```

```
# 建立网络模型
model = keras.Model(inputs=[image_inputs, info_inputs],
                    outputs=[weather_outputs, temp_outputs, humidity_outputs])

# 显示网络模型架构
plot_model(model, to_file='Functional_API_Multi_Input_And_Output_Model.png')
    Image('Functional_API_Multi_Input_And_Output_Model.png')
```

结果如图 2-15 所示。

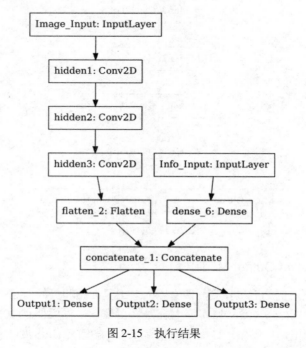

图 2-15　执行结果

2.7　tf.data

2.7.1　tf.data 介绍

从读取数据到数据传入加速设备（GPU 或 TPU）的流程被称为输入管道（Input Pipeline）。输入管道可以分为以下 3 个步骤：

Step 01 提取（Extraction）。

从存储的地方（可以是 SSD、HDD 或远程存储位置）读取数据。

Step 02 转换（Transformation）。

使用 CPU 进行数据预处理，例如对图像进行翻转、裁剪、缩放和正则化等。

Step 03 载入（Loading）。

将转换后的数据加载到机器学习模型的加速设备。

上面这 3 个步骤主要是设备读取数据和 CPU 预处理在消耗时间，如果没有妥善地分工处理，就会造成当 CPU 在准备数据时，GPU 在等待训练数据的产生（GPU 处于空闲状态）；反之，当 GPU 在训练时，CPU 则处于空闲状态，如图 2-16 所示。如此，训练时间就会增加很多。

图 2-16　一般训练时的情况

TensorFlow 提供 tf.data API，可以帮助使用者打造灵活有效的输入管道，轻松处理大量数据、不同数据格式及复杂的转换。而且通过使用 tf.data.Dataset.prefetch，一行指令就可以让生成数据与训练数据同时进行，进而提升训练效率，如图 2-17 所示。

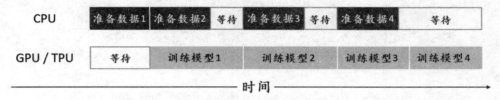

图 2-17　CPU 和 GPU 可同步进行

倘若输入管道的执行时间远比训练时间久，将发生如图 2-18 所示的情况，造成 GPU 或 TPU 加速器无法发挥全部的运算力，通常这种情况可能是读取文件太大或数据预处理太久造成的。

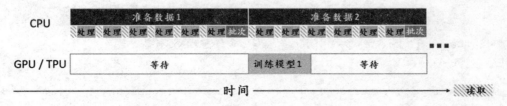

图 2-18　读取与处理时间太长

上述问题可以使用 CPU 多线程来解决，只需在调用 map 方法时加入 num_parallel_calls 设置，即可启用并行处理数据的功能。通常 num_parallel_calls 会设置成计算机的核心数，图 2-19 所示为改善后的工作情况。

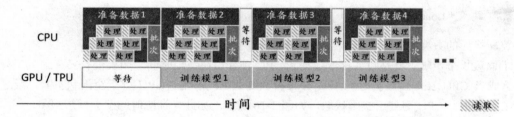

图 2-19 CPU 能以多进程方式准备数据

2.7.2 基本操作

了解了 tf.data 之后，下面将通过几个例子说明如何建立 Dataset、设置 Dataset 和提取数据（注意，下面的程序请依序执行）。

- tf.data.Dataset.from_tensors

使用 from_tensors 建立 Dataset，示例代码如下：

```
dataset = tf.data.Dataset.from_tensors(tf.constant([1, 2, 3, 4, 5, 6, 7,
        8, 9, 10], shape=(10, )))
# 显示 Dataset 信息，形状为(10, )代表一次读取的形状，types 为整数
print(dataset)
```

结果如下：

```
<TensorDataset shapes: (10,), types: tf.int32>
```

- tf.data.Dataset.from_tensor_slices

使用 from_tensor_slices 建立 Dataset，示例代码如下：

```
x_data = tf.data.Dataset.from_tensor_slices(tf.constant([0, 1, 2, 3, 4, 5, 6,
        7, 8, 9]))
# 显示 Dataset 信息，形状为()代表一次读取 1 个数值，types 为整数
print(x_data)

y_data = tf.data.Dataset.from_tensor_slices(tf.constant([0, 2, 4, 6, 8, 10,
12, 14, 16, 18]))
# 显示 Dataset 信息，形状为()代表一次读取 1 个数值，types 为整数
print(y_data)
```

结果如下：

```
<TensorSliceDataset shapes: (), types: tf.int32>
<TensorSliceDataset shapes: (), types: tf.int32>
```

- for loop

读取数据，示例代码如下：

```
for data in dataset:
    print(data)
```

结果如下：

```
tf.Tensor([ 1 2 3 4 5 6 7 8 9 10], shape=(10,), dtype=int32)
```

```
for data1, data2 in zip(x_data, y_data):
    print('x: {}, y: {}'.format(data1, data2))
```

结果如下：

```
x: 0, y: 0
x: 1, y: 2
x: 2, y: 4
x: 3, y: 6
x: 4, y: 8
x: 5, y: 10
x: 6, y: 12
x: 7, y: 14
x: 8, y: 16
x: 9, y: 18
```

- take

通过参数可以指定读取数据的数量，示例代码如下：

```
for data in dataset.take(1):
    print(data)
```

结果如下：

```
tf.Tensor([ 1 2 3 4 5 6 7 8 9 10], shape=(10,), dtype=int32)
```

```
for data1, data2 in zip(x_data.take(5), y_data.take(5)):
    print('x: {}, y: {}'.format(data1, data2))
```

结果如下：

```
x: 0, y: 0
x: 1, y: 2
x: 2, y: 4
x: 3, y: 6
x: 4, y: 8
```

如果指定读取数量超过 Dataset 所含有的数据量，就会读取出 Dataset 中所有的数据。

```
for data1, data2 in zip(x_data.take(12), y_data.take(12)):
    print('x: {}, y: {}'.format(data1, data2))
```

结果如下：

```
x: 0, y: 0
x: 1, y: 2
x: 2, y: 4
x: 3, y: 6
x: 4, y: 8
```

```
x: 5, y: 10
x: 6, y: 12
x: 7, y: 14
x: 8, y: 16
x: 9, y: 18
```

- tf.data.Dataset.zip

将多个 Dataset 打包成一个，示例代码如下：

```
dataset = tf.data.Dataset.zip((x_data, y_data))
print(dataset)
```

结果如下：

```
<ZipDataset shapes: ((), ()), types: (tf.int32, tf.int32)>
```

- map

可以使用 map 来转换数据，示例代码如下：

```
tf.data.Dataset.range(10).map(lambda x: x*2)
```

结果如下：

```
<MapDataset shapes: (), types: tf.int64>
```

- 命名

以字典方式为 elements 的组件命名，示例代码如下：

```
dataset = tf.data.Dataset.zip(
  {"x": tf.data.Dataset.range(10),
   "y": tf.data.Dataset.range(10).map(lambda x: x*2)})
print(dataset)
```

结果如下：

```
<ZipDataset shapes: {y: (), x: ()}, types: {y: tf.int64, x: tf.int64}>
```

```
for data in dataset.take(10):
    print('x: {}, y: {}'.format(data['x'], data['y']))
```

结果如下：

```
x: 0, y: 0
x: 1, y: 2
x: 2, y: 4
x: 3, y: 6
x: 4, y: 8
x: 5, y: 10
x: 6, y: 12
x: 7, y: 14
x: 8, y: 16
x: 9, y: 18
```

- 设置每一批（batch）读取的数据量

```
dataset = tf.data.Dataset.zip(
    {"x": tf.data.Dataset.range(10),
     "y": tf.data.Dataset.range(10).map(lambda x: x*2)}).batch(2)
for data in dataset.take(5):
    print('x: {}, y: {}'.format(data['x'], data['y']))
```

结果如下：

```
x: [0 1], y: [0 2]
x: [2 3], y: [4 6]
x: [4 5], y: [ 8 10]
x: [6 7], y: [12 14]
x: [8 9], y: [16 18]
```

- shuffle

Dataset 数据会被加载到缓冲区中，并从缓冲区中随机选取数据出来，取出数据产生的空位会从新的数据找替补。buffer_size 设置的是缓冲区的大小，最好的设置是大于或等于整个 Dataset 数据的个数。示例代码如下：

```
dataset = dataset.shuffle(10)
for data in dataset.take(5):
    print('x: {}, y: {}'.format(data['x'], data['y']))
```

结果如下：

```
x: [0 1], y: [0 2]
x: [6 7], y: [12 14]
x: [4 5], y: [ 8 10]
x: [8 9], y: [16 18]
x: [2 3], y: [4 6]
```

- repeat

当 Dataset 的数据读取完后，就会读取不到数据，通过设置 repeat(n) 可以重复读取 Dataset n 次，示例代码如下：

```
for data in dataset.take(10):
    print('x: {}, y: {}'.format(data['x'], data['y']))

print('-' * 50)
dataset = dataset.repeat(2)
for data in dataset.take(10):
    print('x: {}, y: {}'.format(data['x'], data['y']))
```

结果如下：

```
x: [0 1], y: [0 2]
x: [6 7], y: [12 14]
x: [4 5], y: [ 8 10]
x: [8 9], y: [16 18]
```

```
x: [2 3], y: [4 6]
--------------------------------------------------
x: [6 7], y: [12 14]
x: [0 1], y: [0 2]
x: [4 5], y: [ 8 10]
x: [8 9], y: [16 18]
x: [2 3], y: [4 6]
x: [8 9], y: [16 18]
x: [4 5], y: [ 8 10]
x: [6 7], y: [12 14]
x: [0 1], y: [0 2]
x: [2 3], y: [4 6]
```

第 3 章

回归问题

学习目标

- 了解神经网络的简史、基本概念及原理
- 了解神经网络权重的更新方法
- 认识 Kaggle 平台
- 运用全连接神经网络完成房价预测系统
- 认识 TensorBoard
- 解决过拟合问题

3.1 深度神经网络

3.1.1 神经网络简史

早在 30 年前，神经网络（Neural Network）[1]-[2] 技术就已经存在，并且盛行过一段时间，不过在 1980—2000 年转而流行支持向量机（Support Vector Machine，SVM）[3-4]，主要的原因是当时能够搭建的神经网络层数不多（1~3 层），导致神经网络所能学习到的特征有限，造成模型性能不佳，这些层数不多的网络被称为浅层神经网络（Shallow Neural Network）。近年来，随着计算机各方面技术的提升，以至于能搭建更庞大的神经网络，因而有了重大的突破，这种更多层数的网络被称为深度神经网络（Deep Neural Network），在性能上远远超越以往的人工智能方法，使神经网络再度受到重视。图 3-1 给出了单隐藏层的浅层神经网络和多隐藏层的深度神经网络的网络架构。

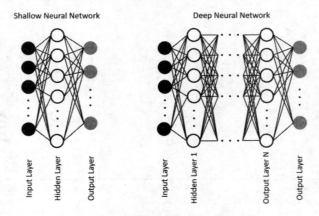

图 3-1　浅层神经网络和深度神经网络

神经网络架构能有突破性发展，归功于以下几个原因：

 运算能力

以往神经网络都是使用 CPU 进行运算的，现在则多使用 GPU 或 TPU 进行运算，执行性能提高了 10~100 倍。这种进展都要归功于 2012 年深度学习之父 Geoffrey Hinton 的学生使用 NVIDIA 的 GPU 训练了 8 层的网络 AlexNet[5]，这个神经网络在当时的图像识别比赛 ImageNet[6] 中以将近 10%的领先优势拿下冠军。至此，大家才意识到 GPU 惊人的运算能力以及深度神经网络的强大。

 数据数量

深度神经网络如果没有足够的训练数据是无法训练起来的，在 20~30 年前，一般不可能拥有大量的数据，原因是当时的存储设备昂贵、容量少以及凭借个人力量难以收集。而现在存储设备价格下降、网络及云端设备盛行，再加上许多开放式的数据库供研究人员使用，故而当时人们所面临的困境不复存在。

 激活函数

激活函数（Activation function）又称为活化函数，以前隐藏层所使用的激活函数大多是 Sigmoid，网络层数太深会导致梯度消失（Vanishing Gradient），使得网络难以更新，因此 X. Glorot 等人在 2011 年提出使用 ReLU[7] 函数，可以有效避免梯度消失问题。

3.1.2　神经网络原理

神经网络可以将其拆为输入层（Input Layer）、隐藏层（Hidden Layer）及输出层（Output Layer），网络中的每一层都有许多神经元（Neuron），各层之间神经元与神经元的连接都是依赖权重（Weight）和偏差（Bias）的，如图 3-2 所示的数学方程式，图中一个圆圈代表一个神经

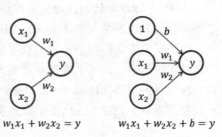

图 3-2　层与层之间的连接

元，左图和右图两边的差别在于有无偏差项。

除了权重和偏差外，网络层的输出还需要加上激活函数，如果没有加上激活函数，那么无论把网络增加多深，网络所能学习到的也只是一个线性方程式，常见的激活函数有 ReLU、Sigmoid、Tanh。激活函数通常会放在每一层的输出后方，如图 3-3 所示。

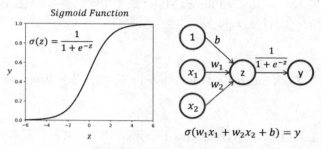

图 3-3　每层的输出后面接上 Sigmoid 激活函数

3.1.3　全连接

在神经网络中，如果每个神经元都与前面一层的所有神经元相连接，就称为全连接（Fully Connected）。图 3-4 所示为三层隐藏层的全连接神经网络，且每一层隐藏层的输出都使用 ReLU 激活函数。本章将实现一个全连接层的房价预测模型。

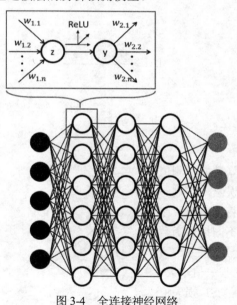

图 3-4　全连接神经网络

3.1.4　损失函数 MSE 和 MAE

损失函数（Loss Function）用来计算模型预测值与预期输出值之间的相似程度，在训练过程中需要将预测值与预期输出值的相似程度最小化。针对不同的问题可以设计合适的损失函数，机器学习中常见的问题有回归问题（Regression Problem）、二分类问题（Binary Classification Problem）和多分类问题（Multiclass Classification Problem）。下面介绍这3种问题以及常用的损失函数。

- 回归问题

通常使用均方误差（Mean Squared Error，MSE）或平均绝对误差（Mean Absolute Error，MAE）来评估预期输出值与预测值的相似程度。

- 二分类问题

通常使用二分类交叉熵（Binary Cross-Entropy，BCE）来评估预期输出值与预测值的相似程度。

- 多分类问题

通常使用多分类交叉熵（Categorical Cross-Entropy，CCE）来评估预期输出值与预测值的相似程度。

本章的主题——房价预测属于回归问题，因此可以使用均方误差和平均绝对误差作为损失函数。下面将介绍这两个损失函数的公式与差异。

➢ 均方误差

$$\mathrm{MSE} = \frac{\sum_{i=1}^{N}(y_i - \hat{y}_i)^2}{N}$$

➢ 平均绝对误差

$$\mathrm{MAE} = \frac{\sum_{i=1}^{N}|y_i - \hat{y}_i|}{N}$$

y：预期输出值。
$\hat{y}$：深度学习模型的预测值。
N：一个批量的数据量。

MSE 与 MAE 都是计算模型预测值（$\hat{y}$）与预期输出值（y）的误差，不过一个将误差值取平方，另一个将误差值取绝对值，相同的预测结果经过不同损失函数的计算会得到不同的损失值，如图 3-5 所示。预测结果为-1~1 时，MAE 损失值较大，而预测值大于 1 或小于-1 时，MSE 损失值较大。采用这两种损失函数会有不同的训练结果，本章范例以 MSE 作为训练的损失函数，读者可以自行更改为 MAE 比较两者的训练差异。

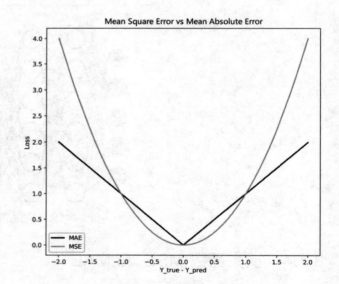

图 3-5　MSE 和 MAE 的损失函数误差比较

3.1.5 神经网络权重更新

梯度下降（Gradient Descent，GD）

神经网络的目的是找到合适的权重与偏差值使损失函数（Loss Function）的损失值（Loss）越来越小，寻找参数的过程称为"学习"，梯度下降是常用的优化算法，通过计算梯度并沿着梯度反方向移动，进而减少网络预测值与样本标注值之间的误差。该算法可以比喻为一个人在山上寻找下山的路，以人当前所处的位置为基准，通过每次寻找周边最陡峭的山路，然后朝着下坡的方向走，反复采用这个方法，每次走一段距离，直到抵达山谷，如图3-6所示。网络参数通过梯度下降得到优化，逐步降低损失值，以找到区域最小的损失值，使网络的预测结果值最接近预期输出值，公式如下：

$$W = W - \eta \frac{\partial L}{\partial W}$$

L：损失函数。
η：学习率。
W：权重。

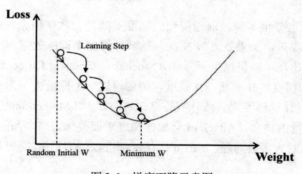

图3-6 梯度下降示意图

事实上，梯度下降每一次的更新并不见得是朝着损失函数的最小值更新，而是朝着当前能减少最多损失函数误差的方向更新，如图3-7所示。因此，当我们训练到损失值无法再降低时，这时所到达的点并非全局最小值（Global Minimum），而是局部最小值（Local Minimum）。

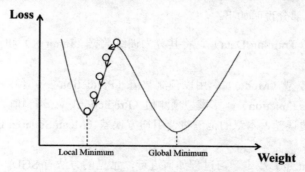

图3-7 权重将会朝着当前能减少最多损失函数误差的方向更新

学习率（Learning Rate）

学习率的大小决定优化算法的优化步伐，即图 3-6 中的 Learning Step。如果学习率过大，每次网络参数更新的步长过大，那么很可能会跃过最佳值（Minimum，最小损失值）并且产生震荡现象，如图 3-8（a）所示；反之，如果学习率过小，那么优化的效率可能过低，经过长时间训练仍无法找到最佳值，如图 3-8（b）所示。

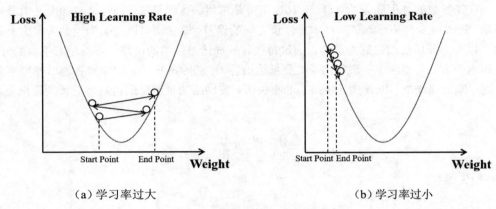

（a）学习率过大　　　　　　　　　　　　（b）学习率过小

图 3-8　不同学习率的优化过程

最后，梯度更新法也分很多种，前面提到的梯度下降是一次使用全部的训练数据计算损失函数的梯度，并更新一次权重，如果要更新 N 次，就要计算整个训练数据 N 遍，这种方法十分花费时间且没有效率。因此出现了随机梯度下降法（Stochastic Gradient Descent，SGD），一次计算一个批量（Batch）数据的梯度值并更新一次权重，例如设置一个批量大小为 64，每一次计算会从训练数据中抽出 64 笔数据计算梯度。另外，还有许多改进方法，如 Momentum[8]、AdaGrad[9]和 Adam[10]等。Momentum 加入动量的概念，AdaGrad 会根据梯度来调整学习率，Adam 可以视为 Momentum 和 AdaGrad 的结合。面对大多数问题，Adam 优化器都可以达到不错的效果。

3.1.6　神经网络训练步骤

图 3-9 所示为神经网络训练示意图，后面的练习都会围绕这张图的 5 部分实现，使读者可以深入浅出地实现具体的流程。

神经网络训练的 5 部分说明如下：

❶ 准备训练数据（Training Data），将其分为训练数据（Input X）和预期输出值的标记答案（Input Y）。

❷ 搭建神经网络模型（Model），用以预测数值（Prediction，ŷ：预测值）。

❸ 损失函数（Loss Function）计算模型预测值（Prediction，ŷ：预测值）与标记答案（Input Y，y：预期输出值）之间的误差，本章的范例将使用均方误差（Mean Squared Error）作为损失函数来训练网络模型。

❹ 优化器（Optimizer）决定学习过程如何进行，常见的方法有 SGD、Adam、RMSprop 等，本范例使用 Adam 优化器。

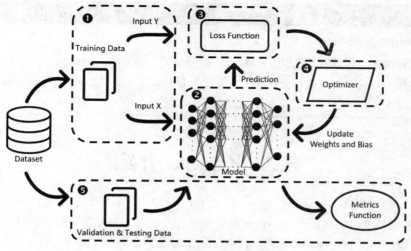

图 3-9　神经网络训练示意图

❺ 准备验证数据（Validation Data）和测试数据（Testing Data）来让网络模型进行预测，并通过评价指标函数（Metrics Function，又称为评价函数）来评价模型的好坏。

> 验证数据：模型训练过程中会调整超参数（网络层数、神经元个数等），以获得最佳模型，在调整过程中利用验证数据的性能作为调整依据。
> 测试数据：大多数比赛会公开训练数据和验证数据，而测试数据会自行保留，作为最后评估模型的主要依据。

说　明

数据通常划分为训练数据、验证数据和测试数据。

测试数据通常于评判一个神经网络算法在真实世界下的正确率和泛化能力强弱的指标。然而，若研究者直接通过测试数据的表现来对网络架构进行修改，则往往能够获得一个在训练数据及测试数据中都有良好表现的网络模型，但应用于真实世界却会表现不佳，此种情况称为元过拟合（Meta-Overfitting）[11]。为了避免研究人员针对测试指标直接进行设计而导致元过拟合，通常会从训练数据中切分出验证数据来作为架构修改的指针，如此测试数据所给出的正确率才会更加符合网络在真实数据下的表现。

说　明

评价指标函数（Metrics Function）

评价指标函数和损失值都用来评估训练模型的好坏程度，有时候评价指标函数和损失值对训练模型的评估结果一致，有时候则不一致。以二分类（Binary Classification）为例，输入两笔数据答案为[1, 0]，并使用两个网络模型进行预测，使用 MAE 计算损失值（计算预测值与标记答案值的误差）和指标值（正确预测数量，预测输出值小于 0.5 视为 0，大于 0.5 则视为 1）：

模型 1：预测输出值为[0.5, 0.5]，得到 MAE 损失值总和为 1.0，正确预测 1 笔数据。
模型 2：预测输出值为[0.7, 0.3]，得到 MAE 损失值总和为 0.6，正确预测 2 笔数据。

> **说 明**
>
> **超参数（Hyperparameter）**
> 超参数是指训练网络模型之前设置的参数，例如批量、学习率、训练周期、网络层数和网络节点数等，这些参数的选择会对训练结果产生影响。第 8 章将会介绍 TensorFlow 辅助调整超参数的工具。

3.2　Kaggle 介绍

Kaggle 是世界著名的竞赛平台，企业和研究学者可以在上面发布数据集或举办竞赛，无论你来自哪里，只要对比赛感兴趣就可以参加，有时候比赛还提供奖金。另外，值得注意的是，Kaggle 除了是竞赛网站之外，也是一个社群平台，参赛者之间可以互相讨论、组队或分享研究成果，而且为了鼓励参赛者多参与讨论，主办单位将奖励分享实验结果并获得最多认可的参赛者。Kaggle 竞赛页面表明当前有高达 19 个竞赛正在进行，如图 3-10 所示。

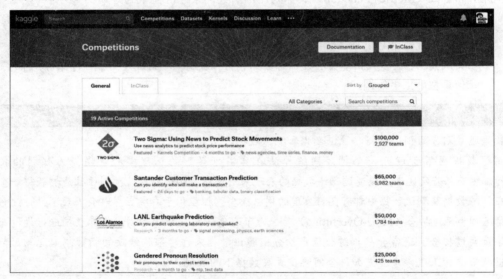

图 3-10　Kaggle 竞赛页面

Kaggle 还有一个很棒的地方是设有 Datasets 专区，里面的数据集都已经过整理并且提供下载，是一个非常好的开源数据集。有些研究员或数据科学家会在数据集下分享自己的研究成果，这使我们学习起来更加如虎添翼。本章的实验数据就是使用 Kaggle 的数据集。

3.3 实验一：房价预测模型

本节的主题为"房价预测"，希望通过神经网络模型来预测房屋的价格。网络模型的输入为房屋的信息，例如面积、楼层或房龄等信息，通过对这些输入信息的分析，网络模型会输出预测的房屋价格。网络模型训练使用均方误差作为损失函数，Adam 作为模型优化器。

3.3.1 数据集介绍

本章的范例使用 Kaggle 的"House Sales in King County, USA"数据集。访问网址：https://www.kaggle.com/harlfoxem/housesalesprediction，并单击 Download 按钮，将压缩文件下载到当前程序代码所保存的目录下，再解压缩这个文件，如图 3-11 所示。

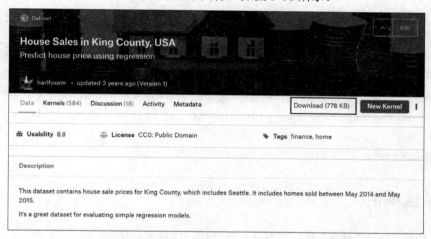

图 3-11　House Sales in King County 数据集下载

此数据集共有 21613 笔房屋数据，每一笔数据有 21 个不同的信息，分别表示以下含义：

- id：房子的标识号。
- date：房屋出售日期。
- price：房屋价格（目标）。
- bedrooms：卧室数量。
- bathrooms：浴室数量。
- sqft_living：居住的面积（平方英尺[①]）。
- sqft_lot：实际的面积（平方英尺）。

---

① 1 平方英尺约等于 0.0929 平方米。

- floors：房屋总楼层数。
- waterfront：海景房。
- view：房屋是否看过。
- condition：整体条件。
- grade：房屋的整体等级（根据 King County 评分系统）。
- sqft_above：除了地下室以外的面积（平方英尺）。
- sqft_basement：地下室的面积（平方英尺）。
- yr_built：房屋建造时间。
- yr_renovated：何时重新装修过（一些未重新装修过或装修记录没被记录到的数值都为 0）。
- zipcode：邮政编码。
- lat：纬度坐标。
- long：经度坐标。
- sqft_living15：2015 年记录的居住面积（可能是翻新的原因，导致 sqft_living15 与 sqft_living 不同）。
- sqft_lot15：2015 年记录的实际面积（可能是翻新的原因，导致 sqft_lot15 与 sqft_lot 不同）。

对于这 21 种不同的信息，我们需要将其分成训练数据（Input X）和预期输出的标记答案（Input Y）两类，分别说明如下：

❶ 训练数据（Input X）：date、bedrooms、bathrooms、sqft_living、sqft_lot、floors、view、waterfront、condition、grade、sqft_above、sqft_basement、yr_built、yr_renovated、lat、zipcode、long、sqft_living15、sqft_lot15。

❷ 预期输出的标记答案（Input Y）：price。

3.3.2 新建项目

建议使用 Jupyter Notebook 执行本章的程序代码，操作流程如下：

Step 01 启动 Jupyter Notebook。

在 Terminal（Ubuntu）或命令提示符（Windows）中输入如下指令：

```
.\tf2\Scripts\activate
jupyter notebook
```

Step 02 新建执行文件。

单击界面右上角的 New 下拉按钮，然后选择所安装的 Python 解释器（在 Jupyter 中都称为 Kernel）来启动它。图 3-12 显示了 3 个不同的 Kernel，分别为：

- Python 3：本地端 Python。
- tf2：虚拟机 Python（TensorFlow-cpu 版本）。
- tf2-gpu：虚拟机 Python（TensorFlow-gpu 版本）。

图 3-12 新建执行文件

Step 03 执行程序代码。

按 Shift + Enter 快捷键来执行单行程序代码,如图 3-13 所示。

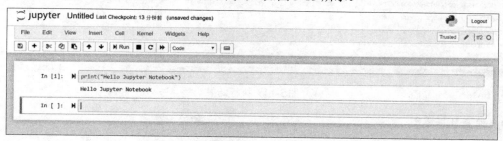

图 3-13 Jupyter 环境界面

接下来,本章后续的程序代码都可在 Jupyter Notebook 上执行。

3.3.3 程序代码

Step 01 导入必要的套件。

```
import os
import numpy as np
import pandas as pd
import tensorflow as tf
import matplotlib.pyplot as plt
from tensorflow import keras
from tensorflow.keras import layers
```

Step 02 数据读取并分析。

- 读取数据

示例代码如下:

```
data = pd.read_csv("kc_house_data.csv")
# 显示数据集的形状,共 21613 笔数据,每一笔数据有 21 种不同信息
data.shape
```

结果如下：

```
(21613, 21)
```

- 显示数据

示例代码如下：

```
# 将显示列数设置为 25，否则会有部分数据无法显示
pd.options.display.max_columns = 25
# head 会显示前 5 行（默认）数据
data.head()
```

结果如图 3-14 所示。

	id	date	price	bedrooms	bathrooms	sqft_living	sqft_lot	floors	waterfront	view	condition	grade	sqft_above	sqft_basem
0	7129300520	20141013T000000	221900.0	3	1.00	1180	5650	1.0	0	0	3	7	1180	
1	6414100192	20141209T000000	538000.0	3	2.25	2570	7242	2.0	0	0	3	7	2170	
2	5631500400	20150225T000000	180000.0	2	1.00	770	10000	1.0	0	0	3	6	770	
3	2487200875	20141209T000000	604000.0	4	3.00	1960	5000	1.0	0	0	5	7	1050	
4	1954400510	20150218T000000	510000.0	3	2.00	1680	8080	1.0	0	0	3	8	1680	

图 3-14　执行结果

- 检查数据类型

数据类型总共有 5 种：字符串（string）、布尔值（boolean）、整数（integer）、浮点数（float）和类别（categorical）。

```
data.dtypes
```

结果如下：

```
id               int64
date             object
price            float64
bedrooms         int64
bathrooms        float64
sqft_living      int64
sqft_lot         int64
floors           float64
waterfront       int64
view             int64
condition        int64
grade            int64
sqft_above       int64
sqft_basement    int64
yr_built         int64
yr_renovated     int64
zipcode          int64
lat              float64
long             float64
sqft_living15    int64
```

```
sqft_lot15          int64
dtype: object
```

Step 03 数据预处理。

- 转换数据类型

因为数据集里的日期（date）数据是字符串（string）类型的，而模型的输入只接收数值类型，所以可以通过以下程序代码将其转为数值类型，并分成年、月、日 3 种数据。

```python
# 将日期拆为年、月、日并转成数值
data['year'] = pd.to_numeric(data['date'].str.slice(0, 4))
data['month'] = pd.to_numeric(data['date'].str.slice(4, 6))
data['day'] = pd.to_numeric(data['date'].str.slice(6, 8))
# 删除没有用的数据，inplace 则是将更新后的数据存回原来的地方
data.drop(['id'], axis="columns", inplace=True)
data.drop(['date'], axis="columns", inplace=True)
data.head()
```

结果如图 3-15 所示。

aterfront	view	condition	grade	sqft_above	sqft_basement	yr_built	yr_renovated	zipcode	lat	long	sqft_living15	sqft_lot15	year	month	day
0	0	3	7	1180	0	1955	0	98178	47.5112	-122.257	1340	5650	2014	10	13
0	0	3	7	2170	400	1951	1991	98125	47.7210	-122.319	1690	7639	2014	12	9
0	0	3	6	770	0	1933	0	98028	47.7379	-122.233	2720	8062	2015	2	25
0	0	5	7	1050	910	1965	0	98136	47.5208	-122.393	1360	5000	2014	12	9
0	0	3	8	1680	0	1987	0	98074	47.6168	-122.045	1800	7503	2015	2	18

图 3-15 执行结果

- 分割数据集

将数据集切割成 3 部分：训练数据（Training Data）、验证数据（Validation Data）和测试数据（Testing Data）。

```python
data_num = data.shape[0]
# 取出一笔与 data 数量相同的随机数索引，主要用于打散数据
indexes = np.random.permutation(data_num)
# 并将随机数索引值分为 train、validation 和 test，这里的划分比例为 6:2:2
train_indexes = indexes[:int(data_num *0.6)]
val_indexes = indexes[int(data_num *0.6):int(data_num *0.8)]
test_indexes = indexes[int(data_num *0.8):]
# 通过索引值从 data 取出训练数据、验证数据和测试数据
train_data = data.loc[train_indexes]
val_data = data.loc[val_indexes]
test_data = data.loc[test_indexes]
```

Step 04 归一化（Normalization）。

归一化的主要作用是将不同的数据缩放到相同的大小，例如，房屋卧室或浴室的数量是 1~5 间，房屋居住的面积为 1500~2500m^2，由于数据量级差异甚大，因此可能会导致网络更加重视数值较大的数据而忽略数值较小的数据。为了解决此问题，我们通常会将输入数据缩放至 0~1 或-1~1，

这个过程就被称为数据归一化（Data Normalization）。

本实验使用标准分数（Standard Score，又称为 z-score）来将数据归一化，经过 z-score 归一化后的数据都会聚集在 0 附近且标准差为 1。

$$x_{norm} = \frac{(x - mean)}{std}$$

```
train_validation_data = pd.concat([train_data, val_data])
mean = train_validation_data.mean()
std = train_validation_data.std()
train_data = (train_data - mean) / std
val_data = (val_data - mean) / std
```

Step 05 建立 NumPy array 格式的训练数据。

```
x_train = np.array(train_data.drop('price', axis='columns'))
y_train = np.array(train_data['price'])
x_val = np.array(val_data.drop('price', axis='columns'))
y_val = np.array(val_data['price'])
```

训练数据共有 12967 笔，每笔有 21 种信息（代表模型输入的数据维度为 21）。

```
x_train.shape
```

结果如下：

```
(12967, 21)
```

Step 06 建立并训练网络模型。

- 搭建全连接网络模型

这里构建 3 层全连接层的网络架构，并且使用 ReLU 作为隐藏层的激活函数，由于需要得到线性输出，因此输出层不使用任何激活函数。

```
# 建立一个 Sequential 类型的模型
model = keras.Sequential(name=' model-1')
# 第 1 层全连接层设为 64 个 unit，将输入形状设置为(21, )，而实际上我们输入的数据形状为
  (batch_size, 21)
model.add(layers.Dense(64, activation='relu', input_shape=(21,)))
# 第 2 层全连接层设为 64 个 unit
model.add(layers.Dense(64, activation='relu'))
# 最后一层全连接层设为 1 个 unit
model.add(layers.Dense(1))
# 显示网络模型架构
model.summary()
```

结果如图 3-16 所示。

```
Model: "model-1"
_____
Layer (type)                 Output Shape              Param #
=================================================================
dense (Dense)                (None, 64)                1408
_____
dense_1 (Dense)              (None, 64)                4160
_____
dense_2 (Dense)              (None, 1)                 65
=================================================================
Total params: 5,633
Trainable params: 5,633
Non-trainable params: 0
```

图 3-16　执行结果

- 设置训练使用的优化器、损失函数和评价指标函数

```
model.compile(keras.optimizers.Adam(0.001),
              loss=keras.losses.MeanSquaredError(),
              metrics=[keras.metrics.MeanAbsoluteError()])
```

- 创建模型存储的目录

```
model_dir = 'lab2-logs/models/'        # 存储模型的位置
os.makedirs(model_dir)                 # 建立存储模型位置的文件夹
```

- 设置回调函数

```
# TensorBoard 回调函数会帮忙记录训练信息，并存成 TensorBoard 的记录文件
log_dir = os.path.join('lab2-logs', 'model-1')
model_cbk = keras.callbacks.TensorBoard(log_dir=log_dir)
# ModelCheckpoint 回调函数帮忙存储网络模型，可以设置只存储最好的模型，monitor 表示被监
测的数据，mode = 'min'则代表监测数据越小越好
model_mckp = keras.callbacks.ModelCheckpoint(model_dir + '/Best-model-1.h5',
                                             monitor='val_mean_absolute_error',
                                             save_best_only=True,
                                             mode='min')
```

- 训练网络模型

```
history = model.fit(x_train, y_train,       # 传入训练数据
                    batch_size=64,           # 批量大小设为 64
                    epochs=300,              # 整个数据集训练 300 遍
                    # 注：1 个 epoch 表示整个训练集中的全部样本训练 1 遍
                    validation_data=(x_val, y_val),    # 验证数据
                    callbacks=[model_cbk, model_mckp]) # Tensorboard回调函数记录训
练过程，ModelCheckpoint 回调函数存储最好的模型
```

结果如图 3-17 所示。

```
Epoch 195/200
12967/12967 [==============================] - 1s 54us/sample - loss: 0.0300 - mean_absolute_error: 0.1283 - val_loss: 0.1763 - val_mean_absolute_error: 0.2283
Epoch 196/200
12967/12967 [==============================] - 1s 55us/sample - loss: 0.0286 - mean_absolute_error: 0.1251 - val_loss: 0.1872 - val_mean_absolute_error: 0.2329
Epoch 197/200
12967/12967 [==============================] - 1s 55us/sample - loss: 0.0417 - mean_absolute_error: 0.1372 - val_loss: 0.1799 - val_mean_absolute_error: 0.2259
Epoch 198/200
12967/12967 [==============================] - 1s 54us/sample - loss: 0.0308 - mean_absolute_error: 0.1265 - val_loss: 0.1763 - val_mean_absolute_error: 0.2394
Epoch 199/200
12967/12967 [==============================] - 1s 57us/sample - loss: 0.0319 - mean_absolute_error: 0.1280 - val_loss: 0.1751 - val_mean_absolute_error: 0.2252
Epoch 200/200
12967/12967 [==============================] - 1s 55us/sample - loss: 0.0348 - mean_absolute_error: 0.1317 - val_loss: 0.1722 - val_mean_absolute_error: 0.2304
```

图 3-17　执行结果

Step 07 训练结果。

- 历史记录（history）

```
history.history.keys()    # 查看 history 存储的信息有哪些
```

结果如下：

```
dict_keys(['loss', 'val_loss', 'val_mean_absolute_error',
          'mean_absolute_error'])
```

- 绘制损失值（loss）的折线图

model.compile 已经将损失函数设为均方误差（Mean Squared Error，MSE），所以 history 记录的 loss 和 val_loss 为 MSE 损失函数计算出来的损失值。

```
plt.plot(history.history['loss'], label='train')
plt.plot(history.history['val_loss'], label='validation')
plt.ylabel('loss')
plt.xlabel('epochs')
plt.legend(loc='upper right')
```

结果如图 3-18 所示。

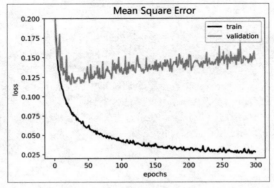

图 3-18　执行结果

- 绘制性能指标（metrics）的折线图

model.compile 已经将评价指标函数设为平均绝对误差（Mean Absolute Error，MAE），所以

网络会计算预测值与答案之间的平均绝对误差，并记录在 history 中（mean_absolute_error 和 val_mean_absolute_error）。

```
plt.plot(history.history['mean_absolute_error'], label='train')
plt.plot(history.history['val_mean_absolute_error'], label='validation')
plt.ylabel('metrics')
plt.xlabel('epochs')
plt.legend(loc='upper right')
```

结果如图 3-19 所示。

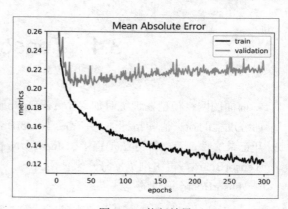

图 3-19　执行结果

- 测试数据的误差百分比

用测试数据预测房屋价格，并与答案计算误差百分比。

```
# 加载模型
model.load_weights('lab2-logs/models/Best-model-1.h5')
# 先将房屋价格取出
y_test = np.array(test_data['price'])
# 归一化数据
test_data = (test_data - mean) / std
# 将输入数据存成 NumPy 格式
x_test = np.array(test_data.drop('price', axis='columns'))
# 预测测试数据
y_pred = model.predict(x_test)
# 将预测结果转换回来(因为训练时的训练目标也经过了归一化)
y_pred = np.reshape(y_pred * std['price'] + mean['price'], y_test.shape)
# 计算平均的误差百分比
percentage_error = np.mean(np.abs(y_test - y_pred)) / np.mean(y_test) * 100
# 显示误差百分比
print("Model_1 Percentage Error: {:.2f}%".format(percentage_error))
```

结果如下：

```
Model_1 Percentage Error: 14.08%
```

> **说　明**
>
> Python format
> 这是一种格式化函数，例如"{:.2f}".format(percentage_error) 只显示到小数点后两位，如果为 "{:.3f}".format(percentage_error)，就显示小数点后 3 位。
> ※ 更多格式可参考：http://www.runoob.com/python/att-string-format.html。

3.4　TensorBoard 介绍

前面的网络模型在训练时加入了 TensorBoard 回调函数（Callback Function），这个函数会记录训练过程的损失值和指标值，并存成 TensorBoard 要求的格式。有两种方法可以使用 TensorBoard 打开记录文件：第一种方法是直接在 Jupyter Notebook 上执行并使用；第二种方法是通过终端机执行 TensorBoard，再由浏览器打开来使用。

TensorBoard 记录的优缺点：

- 优点：训练期间可以实时看到训练的记录信息，不必等到训练全部完成。
- 缺点：训练期间会多次写入记录文件，如果要记录大量信息，就会增加训练时间。

 第一种方法：Jupyter Notebook

```
# 这行指令可以帮助我们直接在 Jupyter Notebook 上显示 TensorBoard
%load_ext tensorboard
# 执行 TensorBoard，并指定记录文件的文件夹为 lab2-logs
# 可以观察每一代训练的损失值 loss 和指标 metrics 的折线图变化
%tensorboard --logdir lab2-logs
```

结果如图 3-20 所示。

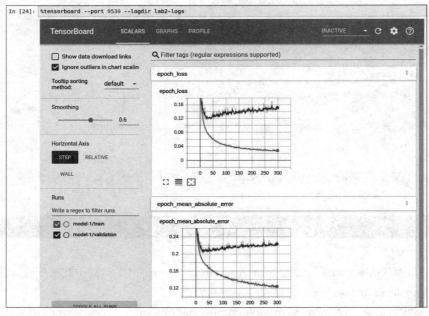

图 3-20　执行结果

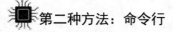 第二种方法：命令行

可以通过网址 http://localhost:6006/ 来访问，--logdir 为 TensorBoard 记录文件的位置。

```
tensorboard --logdir lab2-logs
```

--port 可以设置端口，下面的指令要通过网址 http://localhost:9527/ 来打开，显示结果如图 3-21 所示。

```
tensorboard -port 9527 --logdir lab2-logs
```

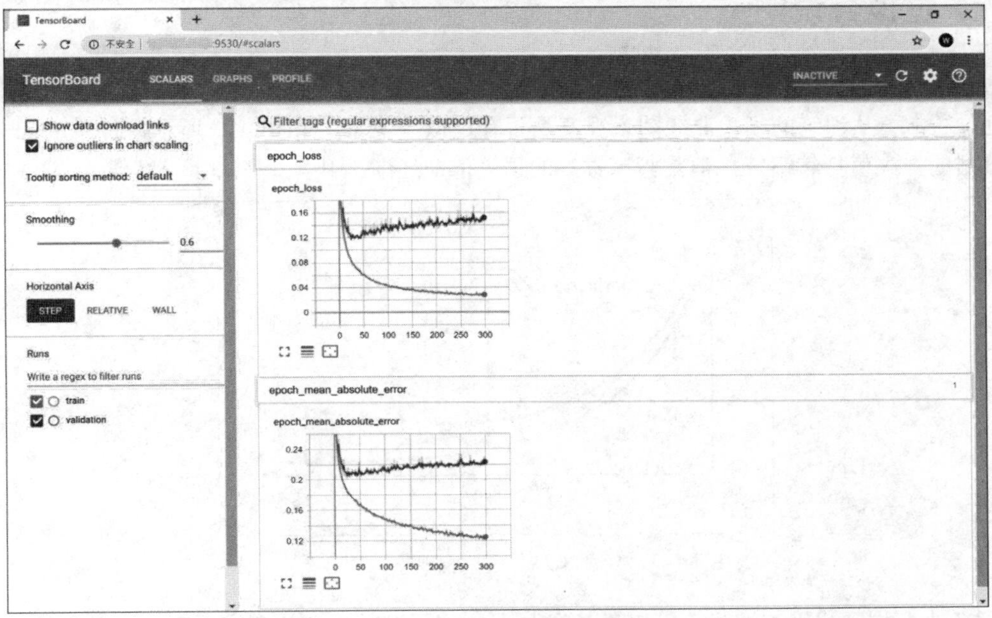

图 3-21　TensorBoard 网页启动示意图

除了记录损失值与指标值外，还有存储模型示意图，如图 3-22 所示，后面的章节会陆续介绍 Distributions、Histograms 和 Images 等。

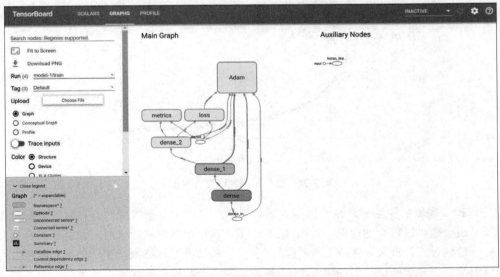

图 3-22　TensorBoard 模型示意图

3.5 实验二：过拟合问题

3.5.1 过拟合说明

过拟合（Overfitting）是指训练的网络模型对验证数据集的性能很差，但对训练数据集的性能却很好。通常会从训练的损失曲线图来观察是否有过拟合现象，如图 3-23 所示，训练一段时间后，训练数据的损失值持续下降，但是验证数据的损失值却逐渐上升。

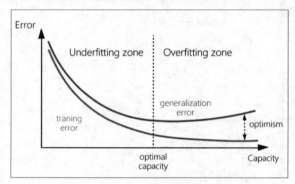

图 3-23　过拟合现象

3.3.3 小节程序代码的训练结果如图 3-24 所示，也可从损失曲线图中观察出过拟合现象。

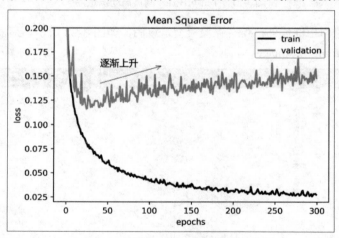

图 3-24　房价预测的过拟合现象

产生过拟合现象的原因通常是"模型太复杂"以及"训练数据太少"，故增加训练数据或简化模型都有机会改善过拟合的问题。如图 3-25 所示，模型参数数量很大时，将能够拟合出复杂的模型，容易造成过拟合现象，图 3-25 左图的灰色分割线是经过训练数据得到的模型结果，可以很完美地将训练数据分成两群，但在图 3-25 右图中，灰色分割线在实际测试数据时会有过拟合的问题，可以发现黑线划分效果比灰线划分效果还要好。

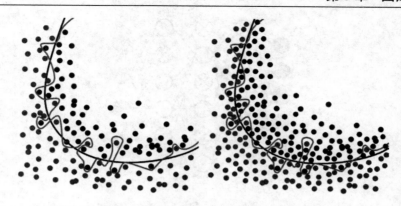

图 3-25　左图为训练数据的分布，右图为实际数据的分布

以下列出 3 种在不增加数据量的条件下防止过拟合的方法。

缩减模型大小

代表模型的参数量减少，参数量少的模型将无法轻易地拟合所有训练数据，因此模型必须学习如何使用有限的参数来学会有效的特征表示。

加入权重正则化（Weights Regularization）

常见的权重正则化方法有 L1 Regularization 和 L2 Regularization（又称作 Weight Decay[12]），核心思想都是通过限制权重的大小解决过拟合问题。例如，如果模型想拟合所有训练数据，就必须依赖复杂的参数，而权重正则化则抵制模型太过于依赖特定的参数。

- L1 Regularization：

$$\text{Loss}_{Total} = \text{Loss}_{MSE} + \lambda \sum_{j=0}^{M} |w_j|$$

- L2 Regularization：

$$\text{Loss}_{Total} = \text{Loss}_{MSE} + \lambda \sum_{j=0}^{M} w_j^2$$

λ：可调整参数，用来控制权重正则化的强度。
w：模型的权重。
M：模型的总参数量。
Loss_{MSE}：代表损失函数为均方误差。
Loss_{Total}：为"损失值"加上"权重正则化"。

加入 Dropout

Dropout[13]的意思是随机失活模型参数，所以 Dropout 也称为随机失活，随机失活的比例可以自行设置，每次训练会随机失活模型中的部分参数，如图 3-26 所示，Dropout 功能仅在训练时启用，在测试时则不使用 Dropout 功能。简单来说，神经网络在每次训练时都使用不同的神经元去学习，如此将可以有效避免神经网络太过于依赖局部特征。

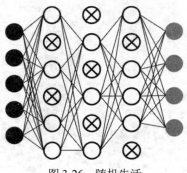

图 3-26　随机失活

3.5.2　程序代码

接续上面的范例,接下来将测试"减少模型大小""加入权重正则化"和"加入 Dropout"3 种方法的训练效果。

- 减少模型大小(隐藏层神经元个数由原来的 64 个降为 16 个)

```
# 建立网络模型
model_2 = keras.Sequential(name='model-2')
# 隐藏层由原来的 64 个降为 16 个
model_2.add(layers.Dense(16, activation='relu', input_shape=(21,)))
# 隐藏层由原来的 64 个降为 16 个
model_2.add(layers.Dense(16, activation='relu'))
model_2.add(layers.Dense(1))

# 设置训练使用的优化器、损失函数和评价指标函数
model_2.compile(keras.optimizers.Adam(0.001),
            loss=keras.losses.MeanSquaredError(),
            metrics=[keras.metrics.MeanAbsoluteError()])

# 设置回调函数
log_dir = os.path.join('lab2-logs', 'model-2')
model_cbk = keras.callbacks.TensorBoard(log_dir=log_dir)
model_mckp = keras.callbacks.ModelCheckpoint(model_dir + '/Best-model-2.h5',
                            monitor='val_mean_absolute_error',
                            save_best_only=True,
                            mode='min')

# 训练网络模型
model_2.fit(x_train, y_train,
        batch_size=64,
        epochs=300,
        validation_data=(x_val, y_val),
        callbacks=[model_cbk, model_mckp])
```

训练结果如图 3-27 所示。

```
Epoch 296/300
12967/12967 [==============================] - 1s 41us/sample - loss: 0.0725 - mean_absolute_error: 0.1754 - val_loss: 0.
1202 - val_mean_absolute_error: 0.2006
Epoch 297/300
12967/12967 [==============================] - 1s 43us/sample - loss: 0.0720 - mean_absolute_error: 0.1760 - val_loss: 0.
1188 - val_mean_absolute_error: 0.2023
Epoch 298/300
12967/12967 [==============================] - 1s 43us/sample - loss: 0.0729 - mean_absolute_error: 0.1765 - val_loss: 0.
1210 - val_mean_absolute_error: 0.2018
Epoch 299/300
12967/12967 [==============================] - 1s 43us/sample - loss: 0.0721 - mean_absolute_error: 0.1758 - val_loss: 0.
1258 - val_mean_absolute_error: 0.2015
Epoch 300/300
12967/12967 [==============================] - 1s 48us/sample - loss: 0.0737 - mean_absolute_error: 0.1769 - val_loss: 0.
1200 - val_mean_absolute_error: 0.2032
<tensorflow.python.keras.callbacks.History at 0x7ff73c7b4940>
```

图 3-27　训练结果

- 加入权重正则化

此模型使用 L2 Regularization 方法，在每一层隐藏层都加上 L2 Regularization，并使用 $\lambda=0.001$ 的正则化强度。

```python
# 建立网络模型
model_3 = keras.Sequential(name='model-3')
# 将此层的所有 w 加上 keras.regularizers.l2 正则化
model_3.add(layers.Dense(64, kernel_regularizer=keras.regularizers
.l2(0.001), activation='relu', input_shape=(21,)))
# 将此层的所有 w 加上 keras.regularizers.l2 正则化
model_3.add(layers.Dense(64,
          kernel_regularizer=keras.regularizers.l2(0.001),
          activation='relu'))
model_3.add(layers.Dense(1))

# 设置训练使用的优化器、损失函数和评价指标函数
model_3.compile(keras.optimizers.Adam(0.001),
          loss=keras.losses.MeanSquaredError(),
          metrics=[keras.metrics.MeanAbsoluteError()])

# 设置回调函数
log_dir = os.path.join('lab2-logs', 'model-3')
model_cbk = keras.callbacks.TensorBoard(log_dir=log_dir)
model_mckp = keras.callbacks.ModelCheckpoint(model_dir + '/Best-model-3.h5',
                                  monitor='val_mean_absolute_error',
                                  save_best_only=True,
                                  mode='min')

# 训练网络模型
model_3.fit(x_train, y_train,
       batch_size=64,
       epochs=300,
       validation_data=(x_val, y_val),
       callbacks=[model_cbk, model_mckp])
```

训练结果如图 3-28 所示。

```
Epoch 296/300
12967/12967 [==============================] - 1s 44us/sample - loss: 0.0687 - mean_absolute_error: 0.1528 - val_loss: 0.
1293 - val_mean_absolute_error: 0.1924
Epoch 297/300
12967/12967 [==============================] - 1s 45us/sample - loss: 0.0744 - mean_absolute_error: 0.1571 - val_loss: 0.
1323 - val_mean_absolute_error: 0.1927
Epoch 298/300
12967/12967 [==============================] - 1s 45us/sample - loss: 0.0756 - mean_absolute_error: 0.1587 - val_loss: 0.
1440 - val_mean_absolute_error: 0.1964
Epoch 299/300
12967/12967 [==============================] - 1s 44us/sample - loss: 0.0717 - mean_absolute_error: 0.1555 - val_loss: 0.
1275 - val_mean_absolute_error: 0.1890
Epoch 300/300
12967/12967 [==============================] - 1s 44us/sample - loss: 0.0691 - mean_absolute_error: 0.1528 - val_loss: 0.
1320 - val_mean_absolute_error: 0.1907
<tensorflow.python.keras.callbacks.History at 0x7ff73c18fc18>
```

图 3-28　训练结果

- 加入 Dropout（随机失活比例设置为 30%）

```python
# 建立网络模型
model_4 = keras.Sequential(name='model-4')
model_4.add(layers.Dense(64, activation='relu', input_shape=(21,)))
# 加入 Dropout 层，并将随机失活比例设置为 30%
model_4.add(layers.Dropout(0.3))
model_4.add(layers.Dense(64, activation='relu'))
# 加入 Dropout 层，并将随机失活比例设置为 30%
model_4.add(layers.Dropout(0.3))
model_4.add(layers.Dense(1))

# 设置训练使用的优化器、损失函数和评价指标函数
model_4.compile(keras.optimizers.Adam(0.001),
            loss=keras.losses.MeanSquaredError(),
            metrics=[keras.metrics.MeanAbsoluteError()])

# 设置回调函数
log_dir = os.path.join('lab2-logs', 'model-4')
model_cbk = keras.callbacks.TensorBoard(log_dir=log_dir)
model_mckp = keras.callbacks.ModelCheckpoint(model_dir + '/Best-model-4.h5',
                                    monitor='val_mean_absolute_error',
                                    save_best_only=True,
                                    mode='min')

# 训练网络模型
model_4.fit(x_train, y_train,
        batch_size=64,
        epochs=300,
        validation_data=(x_val, y_val),
        callbacks=[model_cbk, model_mckp])
```

训练结果如图 3-29 所示。

```
Epoch 296/300
12967/12967 [==============================] - 1s 45us/sample - loss: 0.1047 - mean_absolute_error: 0.2079 - val_loss: 0.
1270 - val_mean_absolute_error: 0.2156
Epoch 297/300
12967/12967 [==============================] - 1s 46us/sample - loss: 0.1056 - mean_absolute_error: 0.2094 - val_loss: 0.
1307 - val_mean_absolute_error: 0.2149
Epoch 298/300
12967/12967 [==============================] - 1s 45us/sample - loss: 0.1154 - mean_absolute_error: 0.2121 - val_loss: 0.
1203 - val_mean_absolute_error: 0.2126
Epoch 299/300
12967/12967 [==============================] - 1s 45us/sample - loss: 0.0971 - mean_absolute_error: 0.2069 - val_loss: 0.
1160 - val_mean_absolute_error: 0.2067
Epoch 300/300
12967/12967 [==============================] - 1s 46us/sample - loss: 0.0996 - mean_absolute_error: 0.2049 - val_loss: 0.
1214 - val_mean_absolute_error: 0.2155
<tensorflow.python.keras.callbacks.History at 0x7ff71c31f438>
```

图 3-29　训练结果

最后，将 3 种模型在测试数据上验证。

- 减少模型大小

```
model_2.load_weights('lab2-logs/models/Best-model-2.h5')
y_pred = model_2.predict(x_test)
y_pred = np.reshape(y_pred * std['price'] + mean['price'], y_test.shape)
percentage_error = np.mean(np.abs(y_test - y_pred)) / np.mean(y_test) * 100
print("Model_2 Percentage Error: {:.2f}%".format(percentage_error))
```

结果如下：

```
Model_2 Percentage Error: 13.15%
```

- 加入权重正则化

```
model_3.load_weights('lab2-logs/models/Best-model-3.h5')
y_pred = model_3.predict(x_test)
y_pred = np.reshape(y_pred * std['price'] + mean['price'], y_test.shape)
percentage_error = np.mean(np.abs(y_test - y_pred)) / np.mean(y_test) * 100
print("Model_3 Percentage Error: {:.2f}%".format(percentage_error))
```

结果如下：

```
Model_3 Percentage Error: 12.89%
```

- 加入 Dropout

```
model_4.load_weights('lab2-logs/models/Best-model-4.h5')
y_pred = model_4.predict(x_test)
y_pred = np.reshape(y_pred * std['price'] + mean['price'], y_test.shape)
percentage_error = np.mean(np.abs(y_test - y_pred)) / np.mean(y_test) * 100
print("Model_4 Percentage Error: {:.2f}%".format(percentage_error))
```

结果如下：

```
Model_4 Percentage Error: 13.33%
```

3.5.3 TensorBoard 数据分析

使用 TensorBoard 观察以及分析以上训练结果。

Step 01 取消左下角所有 Train 记录数据的勾选，只显示 Validation 记录数据。

Model-1/Validation：表示过拟合模型；Model-2/Validation：表示减少参数的模型；Model-3/Validation：代表加入 L2 正则化的模型；Model-4/Validation：代表加入 Dropout 的模型，如图 3-30 所示。

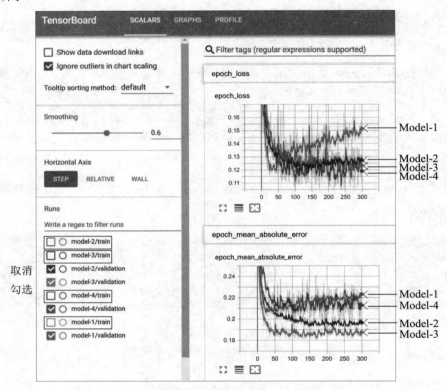

图 3-30 TensorBoard SCALARS 示意图

Step 02 将 Smoothing 比值调到 0，显示不修改原始数据，然后单击折线图下方的缩放图标，将图像缩放至合适的大小，如图 3-31 所示。

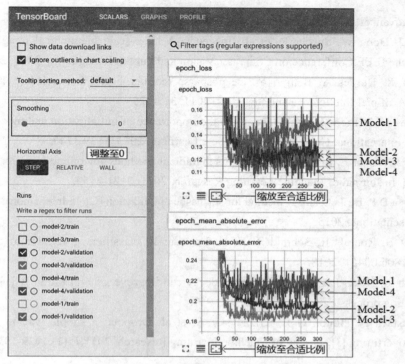

图 3-31　TensorBoard SCALARS 示意图

说　明
调大 Smoothing 比值可以更方便地观察曲线的变动趋势，而将 Smoothing 设置为 0 则更方便找到最低点。

由实验结果证明，这 3 种方法改善了原先模型的过拟合问题，同时也提升了模型的性能，其中 Dropout 方法的损失值最低，而 L2 正则化方法的指标值最低，证明 3 种方法可以有效地改善过拟合问题。

3.6　参考文献

[1] McCulloch W S, Pitts W. A logical calculus of the ideas immanent in nervous activity [J]. The bulletin of mathematical biophysics, 1943,5(4):115-133.

[2] Rosenblatt F. The perceptron: A probabilistic model for information storage and organization in the brain [J]. Psychological Review, 1958,65(6):386-408.

[3] Cortes C, Vapnik V. Support-Vector Network [J]. Machine Learning, 1995, 273-297.

[4] Joachims T. Text categorization with support vector machines: Learning with many relevant features [C]. In European Conference on Machine Learning, 1998, 137-142.

[5] Krizhevsky A, Sutskever I, Hinton G E. Imagenet classification with deep convolutional neural

networks [C]. Advances in Neural Information Processing Systems, 2012, 1097-1105.

[6] Deng J, Dong W, Socher R, et al. Imagenet: A large-scale hierarchical image database [J]. Proceedings of the IEEE Conference on Computer Vision and Pattern Recognition, 2009, 248-255.

[7] Glorot X, Bordes A, Bengio Y. Deep sparse rectifier neural networks[C]. In International Conference on Artificial Intelligence and Statistics, 2011, 315-323.

[8] Sutskever I, Martens J, Dahl G, et al. On the importance of initialization and momentum in deep learning [C]. In International Conference on Machine Learning, 2013, 1139-1147.

[9] Duchi J, Hazan E, Singer Y. Adaptive subgradient methods for online learning and stochastic optimization [J]. In Journal of Machine Learning Research, 2011, 2121-2159.

[10] Kingma D P, Ba J. Adam: A method for stochastic optimization [C]. In International Conference for Learning Representations, 2015.

[11] Recht B, Roelofs R, Schmidt L, et al. Do cifar-10 classifiers generalize to cifar-10? arXiv preprint arXiv:1806.00451, 2018.

[12] Krogh A, Hertz J A. A simple weight decay can improve generalization [C]. In Advances in Neural Information processing systems, 1992, 950-957.

[13] Srivastava N, Hinton G E, Krizhevsky A, et al. Dropout: a simple way to prevent neural networks from overfitting [J]. Journal of Machine Learning Research, 2015,15(1):1929-1958.

第 4 章

二分类问题

学习目标

- 认识机器学习四大类别
- 认识二分类问题
- 了解 Sigmoid 激活函数的使用
- 了解 Binary Cross-Entropy 的使用
- 了解 One-hot Encoding 的使用
- 运用全连接神经网络完成精灵宝可梦对战系统

4.1 机器学习的四大类别

机器学习是一个统称的名词，为人工智能中的一大家族，根据学习方式大致上可以分为 4 类：监督式学习（Supervised Learning）、无监督式学习（Unsupervised Learning）、半监督式学习（Semi-Supervised Learning）和增强式学习（Reinforcement Learning）。下面将介绍每个类别有哪些应用及方法。

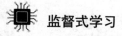
监督式学习

通过已标记（标签）的训练数据集来训练神经网络，标记也可以称作标注（Annotation）。监督式学习的应用非常广泛，除了回归、分类之外，还有目标检测（Object Detection）、图像分割（Image Segmentation）和语音识别（Speech Recognition）等，深度学习的应用几乎都是监督式学习。以下是常见的监督式学习方法：

- 最邻近法（k-Nearest Neighbors，KNN）[1]-[4]。
- 决策树（Decision Trees）[5]-[8]和随机森林（Random Forests）[9]-[11]。
- 支持向量机（Support Vector Machines，SVM）[12]-[15]。
- 深度神经网络（Deep Neural Networks，DNN）[16]-[19]。

无监督式学习

只有训练数据，但训练数据没有标记或标注（或不带标签），常见的任务如聚类（Clustering）和降维（Dimensionality Reduction）。聚类用于量测数据之间的相似度，把性质相同的特征汇聚在一起来实现自动分群。例如，Google 新闻每天都会收集来自全世界的新闻，然后将新闻自动分成十几个群（国际、商业、科技、娱乐和体育等），内容相似的会被分到同一个群。降维用于保持数据的结构以及特征，将数据简化。以下是常见的无监督式学习方法：

- 聚类
  - k-平均算法（k-Means）[20]-[23]。
  - 分级聚类算法（Hierarchical Cluster Analysis，HCA）[24]-[26]。
- 降维
  - 主成分分析（Principal Component Analysis，PCA）[27]-[29]。
  - 核主成分分析（Kernel PCA）[30]-[32]。
  - 自编码（Autoencoder）[33]-[34]。

半监督式学习

通常数据集拥有大量未标记的数据和少量标记的数据，大多数半监督式学习都是监督式学习与无监督式学习的组合。例如，我们将照片上传到相册，相册会自动将人物进行聚类，将看起来相似的归类在一起，而我们只需要给聚类相同的相册贴上标签，之后搜索标签相似的照片便会出现，这是一种监督式与无监督式的结合。

增强式学习

随着 Google DeepMind 团队成功将增强式学习应用在 Atari 游戏、围棋（AlphaGo[35]-[36]）和机器人上，增强式学习开始受到大家的关注与重视。在增强式学习中，学习代理人（Agent）根据接收到的环境状态（State）做出较佳的决策或动作，执行动作（Action）后，会到达下一个状态并获得奖赏（Reward），如图 4-1 所示。例如，在打方块游戏中，将方块消除能得到奖赏；反之，如果死掉的话，就会受到惩罚。增强式学习的宗旨为在一连串的决策过程中最大化期望奖励（Expect Reward）。

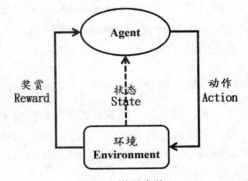

图 4-1　增强式学习

4.2 二分类问题

4.2.1 逻辑回归

首先，逻辑回归（Logistic Regression）并非第 3 章介绍的回归，第 3 章介绍的回归用来预测一个连续的数值，而逻辑回归用来分类，两者有所不同，很多人都会将其搞混。所以当我们听到 Logistic Regression 模型时，指的是分类问题的模型，而听到 Regression 模型，则指的是预测连续数值的模型。

4.2.2 Sigmoid

前面已经提到过 Sigmoid 激活函数，因为梯度消失的缘故，Sigmoid 激活函数并不会用于隐藏层，但是被用于输出层，像二分类的逻辑回归模型，它的输出通常都会加上 Sigmoid 激活函数。Sigmoid 激活函数的输出介于 0 和 1 之间（0 与 1 分别代表两个不同类别），如图 4-2 所示。这个曲线形状很像 S 字母，故又称作"S 函数"。

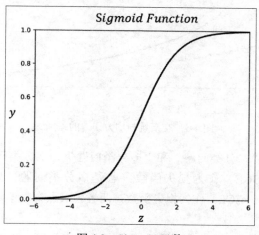

图 4-2　Sigmoid 函数

4.2.3 二分类交叉熵

交叉熵（Cross-Entropy，CE）与均方误差（Mean Squared Error，MSE）是深度学习模型中常见的损失函数。若是回归类问题，则通常使用均方误差当作损失函数，若是分类问题，则通常使用交叉熵。交叉熵的公式如下：

$$CE = -\frac{\sum_{i=1}^{N}\sum_{j=0}^{C} y_{i,j} \log \hat{y}_{i,j}}{N}$$

y：预期输出值。

$\hat{y}$：深度学习模型的预测输出值（预测输出值通常会先经过 Sigmoid 或 Softmax 激活函数）。

C：类别数量。

N：一个批量的数据量。

为什么分类问题不使用均方误差（MSE），而是使用交叉熵（CE）作为损失函数呢？如图4-3所示，当正确分类（$y=1$），MSE 和 CE 的损失值都会为 0，但错误分类（$y=0$）时，MSE 的损失值最大为 1，而 CE 则会爆升至无限大。MSE 的损失值较小，计算的梯度值也会很小，造成网络难以更新。而 CE 的损失值较大，会带来较大的梯度，所以网络模型较容易训练。

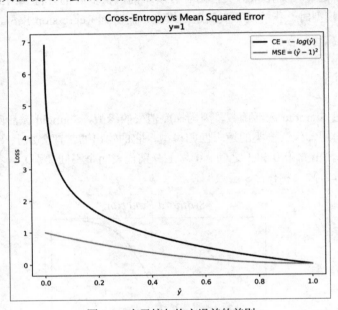

图4-3　交叉熵与均方误差的差别

分类问题又可以分为"二分类问题"和"多分类问题"，二分类问题通常使用二分类交叉熵（Binary Cross-Entropy，BCE）作为损失函数，若是多分类问题，则通常使用多分类交叉熵（Categorical Cross-Entropy，CCE）作为损失函数，两者公式上有些不同，但本质是一样的，都可用"交叉熵"来简称。

首先，本章范例为二分类问题，所以会先介绍二分类交叉熵（BCE）损失函数，二分类交叉熵损失函数为"Sigmoid 激活函数"和"交叉熵损失函数"的组合，所以又称为"Sigmoid 交叉熵"（Sigmoid Cross-Entropy），如图4-4所示。

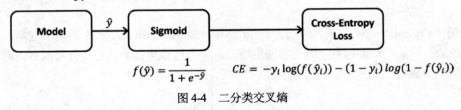

图4-4　二分类交叉熵

所以二分类交叉熵公式如下：

$$BCE = -\frac{\sum_{i=1}^{N}[y_i \log(f(\hat{y}_i)) + (1-y_i)\log(1-f(\hat{y}_i))]}{N}$$

y：预期输出值。
$\hat{y}$：深度学习模型的预测值。
N：一个批量的数据量。
f：Sigmoid 函数。

二分类交叉熵的训练会让预测输出值去逼近 0 或 1。例如，当预期输出值（y）等于 1 时，损失值为$-\log(f(\hat{y}_i))$；当预期输出值（y）等于 0 时，损失值为$-\log(1-f(\hat{y}_i))$，如图 4-5 所示。

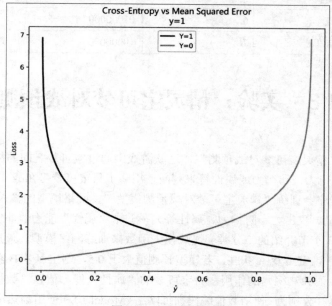

图 4-5　二分类交叉熵

4.2.4　独热编码

独热编码（One-Hot Encoding）是将类别以 0 或 1 表示的方法，如果有 N 个类别，就会以"N-1 个 0"和"1 个 1"来表示。数值表示法和独热编码表示法的比较如表 4-1 所示。"类别的输出与输入"通常会使用独热编码表示，而非数值表示。例如，表 4-1 的人、狗、猫、鸟、……、卡车类别，如果使用数值表示 0、1、2、3、……、7，就会造成"狗比猫更像人（因为 1 比 2 更接近 0）"的潜在关系，但这并不合理，因为这 8 个类别之间是没有连续关系的。独热编码表示法是类别与类别之间完全独立的，所以使用独热编码表示法比数值表示法要好。

表 4-1 数值表示法和独热编码表示法的比较

类别	数值表示	独热编码表示
人	0	00000001
狗	1	00000010
猫	2	00000100
鸟	3	00001000
花	4	00010000
飞机	5	00100000
汽车	6	01000000
卡车	7	10000000

4.3 实验：精灵宝可梦对战预测

本节的主题为"精灵宝可梦对战预测"，这款游戏中通过宝可梦之间的战斗来决定输赢。我们可以依据血量、攻击力、防御力或属性等来判断使用哪一只宝可梦可在战斗中取得胜利。

接下来的范例程序会训练"精灵宝可梦对战预测模型"，网络模型的输入为宝可梦 A 和宝可梦 B 的能力值（血量、攻击力、防御力、属性等），其中"属性"能力值非常适合使用独热编码来表示，并使用 4.2.3 小节介绍的二分类交叉熵损失函数来训练网络模型。最后训练网络模型，若模型预测值小于 0.5，则宝可梦 A 获胜，若模型预测值大于 0.5，则宝可梦 B 获胜。

范例程序为了验证独热编码的作用，将宝可梦的"属性"能力值以"数值表示"方法和"独热编码表示"方法编码，分别为：使用数值编码训练网络（Model 1）和使用独热编码训练网络（Model 2），训练两种不同输入属性表示的网络模型，最后比较数值表示（Model 1）和独热编码表示（Model 2）对网络模型训练正确率的影响。

4.3.1 数据集介绍

本节的范例同样使用 Kaggle 的数据集 Pokemon-Weedle's Cave。打开网址：https://www.kaggle.com/terminus7/pokemon-challenge，并单击 Download 按钮，将压缩文件下载到当前程序代码所保存的目录下，然后解压缩文件，如图 4-6 所示。

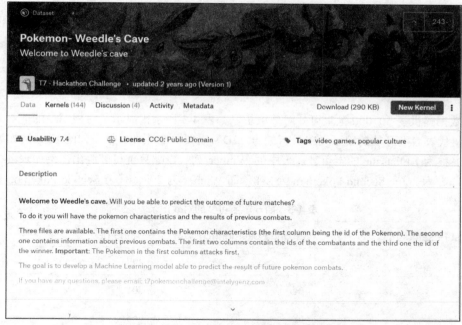

图 4-6 Pokemon-Weedle's Cave 数据集下载

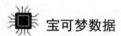 宝可梦数据

表 4-2 显示了前 5 只宝可梦的能力值数据，此数据集会提供 800 只精灵宝可梦的能力值。

表 4-2 前 5 只宝可梦的能力值数据

ID	Name	Type1	Type2	HP	Attack	Defense	Sp・Atk	Sp・Def	Speed	Generation	Legendary
1	Bulbasaur	Grass	Poison	45	49	49	65	65	45	1	False
2	Ivysaur	Grass	Poison	60	62	63	80	80	60	1	False
3	Venusaur	Grass	Poison	80	82	83	100	100	80	1	False
4	MegaVenusaur	Grass	Poison	80	100	123	122	120	80	1	False
5	Charmander	Fire		39	52	43	60	50	65	1	False

Name：宝可梦名字。

Type 1：宝可梦的属性。

Type 2：宝可梦的属性（宝可梦可能拥有两种属性）。

HP：生命值。

Attack：攻击力。

Defense：守备力。

Sp. Atk：特殊攻击力。

Sp. Def：特殊防御力。

Speed：速度。
Generation：进化阶段。
Legendary：神兽（1=神兽，0=一般）。

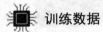 训练数据

训练数据提供了 50 000 笔宝可梦对战记录，如表 4-3 所示，显示了对战记录的前 5 笔数据。表中的数值代表着宝可梦的编号（ID），编号（ID）需要对照表 4-2 第一列的 ID，例如编号 1 对照表 4-2 名称为 Bulbasaur 的宝可梦。表中的 Winner 列代表哪只宝可梦胜出，例如表 4-3 第一行 First Pokemon 为 266，Second Pokemon 为 298，而 Winner 为 298，代表 Second Pokemon 获胜。

表 4-3 前 5 笔宝可梦的对战记录

First Pokemon(ID)	Second Pokemon(ID)	Winner(ID)
266	298	298
702	701	701
191	668	668
237	683	683
151	231	151

如图 4-7 所示，First Pokemon（ID：5）对应 Charmander 宝可梦，Second Pokemon（ID：1）对应 Bulbasaur 宝可梦，（ID：5）对应 Charmander 宝可梦。

宝可梦其中一笔对战记录

First Pokemon(ID)	Second Pokemon(ID)	Winner(ID)
5	1	5

前五只宝可梦的能力数据

ID	Name	Type1	Type2	HP	Attack	Defense	Sp Atk	Sp Def	Speed	Generation	Legendary
1	Bulbasaur	Grass	Poison	45	49	49	65	65	45	1	False
2	Ivysaur	Grass	Poison	60	62	63	80	80	60	1	False
3	Venusaur	Grass	Poison	80	82	83	100	100	80	1	False
4	MegaVenusaur	Grass	Poison	80	100	123	122	120	80	1	False
5	Charmander	Fire		39	52	43	60	50	65	1	False

图 4-7 "对战记录"与"宝可梦数据"的对应关系

因此，我们需要将数据集分成输入训练数据（Input X）和预期输出的标记答案（Input Y）两类，分别说明如下：

训练数据（Input X）：First Pokemon（Type 1、Type 2、HP、Attack、Defense、Sp. Atk、Sp. Def、Speed、Generation、Legendary）和 Second Pokemon（Type 1、Type 2、HP、Attack、Defense、Sp. Atk、Sp. Def、Speed、Generation、Legendary）。

预期输出的标记答案（Input Y）：First Pokemon 或 Second Pokemon，输出 0 表示 First Pokemon

胜出，输出 1 表示 Second Pokemon 胜出。

> **说　明**
>
> 宝可梦数据的属性（Type 1 和 Type 2）总共有 18 种，可以使用 0~17 来表示每种不同的属性。但是，属性与属性之间没有连续关系（说明可查看 4.2.4 小节），所以使用独热编码的方法表示属性较为理想。

4.3.2 新建项目

建议使用 Jupyter Notebook 来执行本小节的程序代码，操作流程如下：

Step 01 启动 Jupyter Notebook。

在 Terminal（Ubuntu）或命令提示符（Windows）中输入如下指令：

```
jupyter notebook
```

Step 02 新建执行文件。

单击右上角的 New 下拉按钮，然后单击所安装的 Python 解释器（在 Jupyter 中都称为 Kernel）来启动它，如图 4-8 所示，显示了 3 个不同的 Kernel：

- Python 3：本地端 Python。
- tf2：虚拟机 Python（TensorFlow-cpu 版本）。
- tf2-gpu：虚拟机 Python（TensorFlow-gpu 版本）。

图 4-8　新建执行文件

Step 03 执行程序代码。

按 Shift + Enter 快捷键执行单行程序代码，如图 4-9 所示。

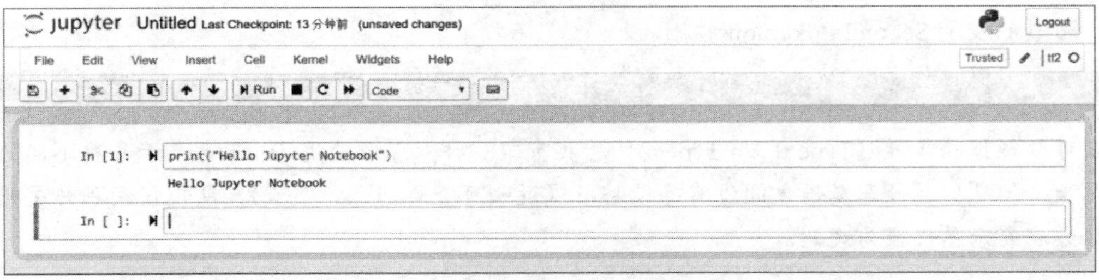

图 4-9 Jupyter 环境界面

接下来,本章后续的程序代码都可在 Jupyter Notebook 上执行。

4.3.3 程序代码

Step 01 导入必要的套件。

```
import os
import numpy as np
import pandas as pd
import tensorflow as tf
import matplotlib.pyplot as plt
from tensorflow import keras
from tensorflow.keras import layers
```

Step 02 读取数据并分析。

- 读取 800 只宝可梦数据

```
pokemon_df = pd.read_csv('./pokemon.csv')
pokemon_df.head()
```

结果如图 4-10 所示。

	#	Name	Type 1	Type 2	HP	Attack	Defense	Sp. Atk	Sp. Def	Speed	Generation	Legendary
0	1	Bulbasaur	Grass	Poison	45	49	49	65	65	45	1	False
1	2	Ivysaur	Grass	Poison	60	62	63	80	80	60	1	False
2	3	Venusaur	Grass	Poison	80	82	83	100	100	80	1	False
3	4	Mega Venusaur	Grass	Poison	80	100	123	122	120	80	1	False
4	5	Charmander	Fire	NaN	39	52	43	60	50	65	1	False

图 4-10 执行结果

- 将"#"数据设置为索引值

```
pokemon_df= pokemon_df.set_index("#")
pokemon_df.head()
```

结果如图 4-11 所示。

#	Name	Type 1	Type 2	HP	Attack	Defense	Sp. Atk	Sp. Def	Speed	Generation	Legendary
1	Bulbasaur	Grass	Poison	45	49	49	65	65	45	1	False
2	Ivysaur	Grass	Poison	60	62	63	80	80	60	1	False
3	Venusaur	Grass	Poison	80	82	83	100	100	80	1	False
4	Mega Venusaur	Grass	Poison	80	100	123	122	120	80	1	False
5	Charmander	Fire	NaN	39	52	43	60	50	65	1	False

图 4-11　执行结果

- 读取宝可梦对战数据

```
combats_df = pd.read_csv('./combats.csv')
combats_df.head()
```

结果如图 4-12 所示。

	First_pokemon	Second_pokemon	Winner
0	266	298	298
1	702	701	701
2	191	668	668
3	237	683	683
4	151	231	151

图 4-12　执行结果

Step 03 读取到缺失值的处理方式和填充数据到缺失值。

- 检查宝可梦数据是否有缺失数据

宝可梦数据总共收录了 800 只宝可梦，但从输出结果会发现 Name 和 Type 2 都有缺失数据。

> ➤ Name 数据可能有原来数据集的缺失，不过并不影响训练，因为训练时并不会使用到名称。
> ➤ Type 2（宝可梦的第二属性值）缺失数据是因为有些宝可梦没有第二属性，因此需要填充数据到缺失的字段中，否则数据处理会有影响。

```
pokemon_df.info()
```

结果如下：

```
<class 'pandas.core.frame.DataFrame'>
Int64Index: 800 entries, 1 to 800
Data columns (total 11 columns):
Name        799 non-null object      ── 缺少了 1 笔数据
Type 1      800 non-null object
Type 2      414 non-null object      ── 缺少了 386 笔数据
HP          800 non-null int64
Attack      800 non-null int64
Defense     800 non-null int64
Sp. Atk     800 non-null int64
```

```
Sp. Def        800 non-null int64
Speed          800 non-null int64
Generation     800 non-null int64
Legendary      800 non-null bool
dtypes: bool(1), int64(7), object(3)
memory usage: 69.5+ KB
```

- 查看 Type 2 每个类别的数量

通过传入参数 dropna=False 可以将缺失数据（NaN）也考虑进去，NaN 代表宝可梦并没有第二种属性。

```
pokemon_df["Type 2"].value_counts(dropna=False)
```

结果如下：

```
NaN       386    ———— 代表没有第二属性的宝可梦有 386 只
Flying     97
Ground     35
Poison     34
Psychic    33
Fighting   26
Grass      25
Fairy      23
Steel      22
Dark       20
Dragon     18
Water      14
Rock       14
Ice        14
Ghost      14
Fire       12
Electric    6
Normal      4
Bug         3
Name: Type 2, dtype: int64
```

- 填充缺失数据

使用 empty 将缺失的字段填上：

```
pokemon_df["Type 2"].fillna('empty',inplace=True)
pokemon_df["Type 2"].value_counts()
```

结果如下：

```
empty     386    ———— NaN → empty
Flying     97
Ground     35
Poison     34
Psychic    33
Fighting   26
Grass      25
```

```
Fairy        23
Steel        22
Dark         20
Dragon       18
Water        14
Rock         14
Ice          14
Ghost        14
Fire         12
Electric      6
Normal        4
Bug           3
Name: Type 2, dtype: int64
```

Step 04 数据预处理。

- 检查数据类型

其中，Type 1、Type 2 和 Legendary 都是要输入网络模型的数据，但是三者的数据类型都无法直接输入模型中，所以需要经过数据转换。

```
print(combats_df.dtypes)
print('-' * 30)
print(pokemon_df.dtypes)
```

结果如下：

```
First_pokemon     int64
Second_pokemon    int64
Winner            int64
dtype:            object
------------------------------
Name          object
Type 1        object
Type 2        object
HP             int64
Attack         int64
Defense        int64
Sp. Atk        int64
Sp. Def        int64
Speed          int64
Generation     int64
Legendary       bool
dtype: object
```

- 转换数据格式

  ➢ Type 1、Type 2：将两个信息从 object 类型转成 category（类别）类型，这种类型下有许多函数可以使用，之后会通过 cat.codes 方法转成数值表示。

  ➢ Legendary：将 Legendary 信息从 bool（布尔）类型转成 int（整数）类型，数据的表示法会从（False, True）变成（0, 1）。

```
# 将 Type 1 转成类别类型
pokemon_df['Type 1'] = pokemon_df['Type 1'].astype('category')
# 将 Type 2 转成类别类型
pokemon_df['Type 2'] = pokemon_df['Type 2'].astype('category')
# 将 Legendary 转成整数类型
pokemon_df['Legendary'] = pokemon_df['Legendary'].astype('int')
pokemon_df.dtypes
```

结果如下：

```
Name           object
Type 1         category
Type 2         category
HP             int64
Attack         int64
Defense        int64
Sp. Atk        int64
Sp. Def        int64
Speed          int64
Generation     int64
Legendary      int64
dtype: object
```

- 将宝可梦的 Type 1 和 Type 2 转换为独热编码

使用 Pandas 的 get_dummies 函数获取 Type 1（宝可梦第一种属性）的独热编码：

```
df_type1_one_hot = pd.get_dummies(pokemon_df['Type 1'])
df_type1_one_hot.head()
```

结果如图 4-13 所示。

#	Bug	Dark	Dragon	Electric	Fairy	Fighting	Fire	Flying	Ghost	Grass	Ground	Ice	Normal	Poison	Psychic	Rock	Steel	Water
1	0	0	0	0	0	0	0	0	0	1	0	0	0	0	0	0	0	0
2	0	0	0	0	0	0	0	0	0	1	0	0	0	0	0	0	0	0
3	0	0	0	0	0	0	0	0	0	1	0	0	0	0	0	0	0	0
4	0	0	0	0	0	0	0	0	0	1	0	0	0	0	0	0	0	0
5	0	0	0	0	0	1	0	0	0	0	0	0	0	0	0	0	0	0

图 4-13 执行结果

使用 Pandas 的 get_dummies 函数获取 Type 2（宝可梦第二种属性）的独热编码：

```
df_type2_one_hot = pd.get_dummies(pokemon_df['Type 2'])
df_type2_one_hot.head()
```

结果如图 4-14 所示。

	Bug	Dark	Dragon	Electric	Fairy	Fighting	Fire	Flying	Ghost	Grass	Ground	Ice	Normal	Poison	Psychic	Rock	Steel	Water	empty
#																			
1	0	0	0	0	0	0	0	0	0	0	0	0	0	1	0	0	0	0	0
2	0	0	0	0	0	0	0	0	0	0	0	0	0	1	0	0	0	0	0
3	0	0	0	0	0	0	0	0	0	0	0	0	0	1	0	0	0	0	0
4	0	0	0	0	0	0	0	0	0	0	0	0	0	1	0	0	0	0	0
5	0	0	0	0	0	0	0	0	0	0	0	0	0	0	0	0	0	0	1

图 4-14　执行结果

将两组独热编码合并回数据集：

```
# 将 Type 1 和 Type 2 两个独热编码合并
combine_df_one_hot = df_type1_one_hot.add(df_type2_one_hot,
                                          fill_value=0).astype('int64')
# 将显示列数设置为 30，否则会有部分数据无法显示
pd.options.display.max_columns = 30
# 将 combine_df (Type1 和 Type2) 与数据集合并
pokemon_df = pokemon_df.join(combine_df_one_hot)
pokemon_df.head()
```

结果如图 4-15 所示。

	Name	Type 1	Type 2	HP	Attack	Defense	Sp. Atk	Sp. Def	Speed	Generation	Legendary	Bug	Dark	Dragon	Electric	Fairy	Fighting	Fire	Flying
#																			
1	Bulbasaur	Grass	Poison	45	49	49	65	65	45	1	0	0	0	0	0	0	0	0	0
2	Ivysaur	Grass	Poison	60	62	63	80	80	60	1	0	0	0	0	0	0	0	0	0
3	Venusaur	Grass	Poison	80	82	83	100	100	80	1	0	0	0	0	0	0	0	0	0
4	Mega Venusaur	Grass	Poison	80	100	123	122	120	80	1	0	0	0	0	0	0	0	0	0
5	Charmander	Fire	empty	39	52	43	60	50	65	1	0	0	0	0	0	0	0	1	0

图 4-15　执行结果

- 将宝可梦属性转为数值表示（0, 1, 2, …, 18）

通过 cat.categories 查询类别的标签：

```
dict(enumerate(pokemon_df['Type 2'].cat.categories))
```

结果如下：

```
{0:     'Bug',
 1:     'Dark',
 2:     'Dragon',
 3:     'Electric',
 4:     'Fairy',
 5:     'Fighting',
 6:     'Fire',
 7:     'Flying',
 8:     'Ghost',
 9:     'Grass',
 10:    'Ground',
 11:    'Ice',
 12:    'Normal',
```

```
13:     'Poison',
14:     'Psychic',
15:     'Rock',
16:     'Steel',
17:     'Water',
18:     'empty'}
```

通过 cat.codes 可以获取类别的编码值：

```
pokemon_df['Type 2'].cat.codes.head(10)
```

结果如下：

```
1     13
2     13
3     13
4     13
5     18
6     18
7     7
8     2
9     7
10    18
dtype: int8
```

用数值表示（0, 1, 2,…, 18）取代原来的标签值：

```
pokemon_df['Type 1'] = pokemon_df['Type 1'].cat.codes
pokemon_df['Type 2'] = pokemon_df['Type 2'].cat.codes
pokemon_df.head()
```

结果如图 4-16 所示。

#	Name	Type 1	Type 2	HP	Attack	Defense	Sp. Atk	Sp. Def	Speed	Generation	Legendary	Bug	Dark	Dragon	Electric	Fairy	Fighting	Fire	Flying	Gl
1	Bulbasaur	9	13	45	49	49	65	65	45	1	0	0	0	0	0	0	0	0	0	0
2	Ivysaur	9	13	60	62	63	80	80	60	1	0	0	0	0	0	0	0	0	0	0
3	Venusaur	9	13	80	82	83	100	100	80	1	0	0	0	0	0	0	0	0	0	0
4	Mega Venusaur	9	13	80	100	123	122	120	80	1	0	0	0	0	0	0	0	0	0	0
5	Charmander	6	18	39	52	43	60	50	65	1	0	0	0	0	0	0	0	0	1	0

图 4-16　执行结果

- 将没有使用到的数据剔除

```
pokemon_df.drop('Name', axis='columns', inplace=True)
pokemon_df.head()
```

结果如图 4-17 所示。

#	Type 1	Type 2	HP	Attack	Defense	Sp. Atk	Sp. Def	Speed	Generation	Legendary	Bug	Dark	Dragon	Electric	Fairy	Fighting	Fire	Flying	Ghost	Grass
1	9	13	45	49	49	65	65	45	1	0	0	0	0	0	0	0	0	0	0	1
2	9	13	60	62	63	80	80	60	1	0	0	0	0	0	0	0	0	0	0	1
3	9	13	80	82	83	100	100	80	1	0	0	0	0	0	0	0	0	0	0	1
4	9	13	80	100	123	122	120	80	1	0	0	0	0	0	0	0	0	0	0	1
5	6	18	39	52	43	60	50	65	1	0	0	0	0	0	0	0	1	0	0	0

图 4-17　执行结果

- 将宝可梦对战数据中胜利方的表示改为 0 与 1

```
# apply 方法第一个参数为自定义函数，axis 为 columns 的话会将数据一行一行放入函数中
# 进行处理，最后将所有结果组合成一个数据结构返回
combats_df['Winner'] = combats_df.apply(lambda x: 0 
                        if x.Winner == x.First_pokemon else 1,
                        axis='columns')
combats_df.head()
```

结果如图 4-18 所示。

	First_pokemon	Second_pokemon	Winner
0	266	298	1
1	702	701	1
2	191	668	1
3	237	683	1
4	151	231	0

图 4-18　执行结果

Step 05 数据集分割。

将数据集分为 3 部分，分别为：

- 训练数据（Training Data）。
- 验证数据（Validation Data）。
- 测试数据（Testing Data）。

```
data_num = combats_df.shape[0]
# 取出一笔与 data 数量相同的随机数索引，主要用于打乱数据
indexes = np.random.permutation(data_num)
# 并将随机数索引值分为 train、validation 和 test，这里划分比例为 6:2:2
train_indexes = indexes[:int(data_num *0.6)]
val_indexes = indexes[int(data_num *0.6):int(data_num *0.8)]
test_indexes = indexes[int(data_num *0.8):]
train_data = combats_df.loc[train_indexes]
val_data = combats_df.loc[val_indexes]
test_data = combats_df.loc[test_indexes]
```

Step 06 归一化（Normalization）。

- 将数值表示的属性除以 19（因为加上 empty 共有 19 种属性），把数值缩放到 0~1

```
pokemon_df['Type 1'] = pokemon_df['Type 1'] / 19
pokemon_df['Type 2'] = pokemon_df['Type 2'] / 19
```

- 使用 Standard Score 将生命值、攻击力和防御力等数值归一化

```
mean = pokemon_df.loc[:, 'HP':'Generation'].mean()
std = pokemon_df.loc[:, 'HP':'Generation'].std()
pokemon_df.loc[:,'HP':'Generation'] = (pokemon_df.loc[:,'HP':'Generation']
-mean)/std
pokemon_df.head()
```

结果如图 4-19 所示。

#	Type 1	Type 2	HP	Attack	Defense	Sp. Atk	Sp. Def	Speed	Generation	Legendary	Bug	Dark	Dragon	Electric	Fairy	Fightin
1	0.473684	0.684211	-0.950032	-0.924328	-0.796655	-0.238981	-0.248033	-0.801002	-1.398762	0	0	0	0	0	0	
2	0.473684	0.684211	-0.362595	-0.523803	-0.347700	0.219422	0.290974	-0.284837	-1.398762	0	0	0	0	0	0	
3	0.473684	0.684211	0.420654	0.092390	0.293665	0.830626	1.009651	0.403383	-1.398762	0	0	0	0	0	0	
4	0.473684	0.684211	0.420654	0.646964	1.576395	1.502951	1.728328	0.403383	-1.398762	0	0	0	0	0	0	
5	0.315789	0.947368	-1.185007	-0.831899	-0.989065	-0.391782	-0.787041	-0.112782	-1.398762	0	0	0	0	0	0	

图 4-19 执行结果

Step 07 建立 NumPy array 格式的训练数据。

- 准备对战数据中每个宝可梦对应能力值的索引

```
x_train_index = np.array(train_data.drop('Winner', axis='columns'))
x_val_index = np.array(val_data.drop('Winner', axis='columns'))
x_test_index = np.array(test_data.drop('Winner', axis='columns'))
print(x_train_index)
```

结果如下：

```
[[115 674]
 [658 549]
 [434  87]
 ...
 [732 607]
 [239 608]
 [298 742]]
```

- 准备训练目标

```
y_train = np.array(train_data['Winner'])
y_val = np.array(val_data['Winner'])
y_test = np.array(test_data['Winner'])
```

- 准备两种不同的输入数据

第一种：宝可梦的属性为数值表示。

```
# 获取宝可梦的能力值
```

```
pokemon_data_normal = np.array(pokemon_df.loc[:, : 'Legendary'])
print(pokemon_data_normal.shape)
# 通过前面准备的索引产生输入数据
x_train_normal = pokemon_data_normal[x_train_index -1].reshape((-1, 20))
x_val_normal = pokemon_data_normal[x_val_index -1].reshape((-1, 20))
x_test_normal = pokemon_data_normal[x_test_index -1].reshape((-1, 20))
print(x_train_normal.shape)
```

结果如下:

```
(800, 10)
(30000, 20)
```

第二种: 宝可梦的属性为独热编码表示。

```
# 获取宝可梦的能力值
pokemon_data_one_hot = np.array(pokemon_df.loc[:, 'HP':])
print(pokemon_data_one_hot.shape)
# 通过前面准备的索引产生输入数据
x_train_one_hot = pokemon_data_one_hot[x_train_index -1].reshape((-1, 54))
x_val_one_hot = pokemon_data_one_hot[x_val_index -1].reshape((-1, 54))
x_test_one_hot = pokemon_data_one_hot[x_test_index -1].reshape((-1, 54))
print(x_train_one_hot.shape)
```

结果如下:

```
(800, 27)
(30000, 54)
```

Step 08 使用数值编码训练网络 (Model 1)。

- 建立网络模型

本小节模型架构参考 3.5.2 小节 "加入 Dropout" 中的模型架构,建立 4 层的全连接隐藏层,并且每一层的输出都会接上 Dropout 层 (随机失活比例设置为 30%)。网络模型的输入为两只宝可梦的数据,每只宝可梦有 10 种不同的数据,所以两只宝可梦输入大小为(20,)。

```
inputs = keras.Input(shape=(20, ))
x = layers.Dense(64, activation='relu')(inputs)
x = layers.Dropout(0.3)(x)
x = layers.Dense(64, activation='relu')(x)
x = layers.Dropout(0.3)(x)
x = layers.Dense(64, activation='relu')(x)
x = layers.Dropout(0.3)(x)
x = layers.Dense(16, activation='relu')(x)
x = layers.Dropout(0.3)(x)
outputs = layers.Dense(1, activation='sigmoid')(x)

model_1 = keras.Model(inputs, outputs, name='model-1')
# 显示网络架构
model_1.summary()
```

结果如图 4-20 所示。

```
Model: "model-1"
_____
Layer (type)                 Output Shape              Param #
=================================================================
input_1 (InputLayer)         [(None, 20)]              0
_____
dense (Dense)                (None, 64)                1344
_____
dropout (Dropout)            (None, 64)                0
_____
dense_1 (Dense)              (None, 64)                4160
_____
dropout_1 (Dropout)          (None, 64)                0
_____
dense_2 (Dense)              (None, 64)                4160
_____
dropout_2 (Dropout)          (None, 64)                0
_____
dense_3 (Dense)              (None, 16)                1040
_____
dropout_3 (Dropout)          (None, 16)                0
_____
dense_4 (Dense)              (None, 1)                 17
=================================================================
Total params: 10,721
Trainable params: 10,721
Non-trainable params: 0
```

图 4-20 执行结果

- 设置训练使用的优化器、损失函数和评价指标函数

```
model_1.compile(keras.optimizers.Adam(),
            loss=keras.losses.BinaryCrossentropy(),
            metrics=[keras.metrics.BinaryAccuracy()])
```

- 创建模型存储的目录

```
model_dir = 'lab3-logs/models'
os.makedirs(model_dir)
```

- 设置回调函数

```
# 将训练记录存成 TensorBoard 的记录文件
log_dir = os.path.join('lab3-logs', 'model-1')
model_cbk = keras.callbacks.TensorBoard(log_dir=log_dir)
# 存储最好的网络模型权重
model_mckp = keras.callbacks.ModelCheckpoint(model_dir + '/Best-model-1.h5',
                                    monitor='val_binary_accuracy',
                                    save_best_only=True,
                                    mode='max')
```

- 训练网络模型

```
history_1 = model_1.fit(x_train_normal, y_train,
                    batch_size=64,
                    epochs=200,
                    validation_data=(x_val_normal, y_val),
                    callbacks=[model_cbk, model_mckp])
```

结果如图 4-21 所示。

```
Epoch 195/200
30000/30000 [==============================] - 2s 51us/sample - loss: 0.1413 - binary_accuracy: 0.9506 - val_loss: 0.1627 - val_binary_accuracy: 0.9466
Epoch 196/200
30000/30000 [==============================] - 2s 52us/sample - loss: 0.1451 - binary_accuracy: 0.9489 - val_loss: 0.1639 - val_binary_accuracy: 0.9472
Epoch 197/200
30000/30000 [==============================] - 2s 50us/sample - loss: 0.1422 - binary_accuracy: 0.9503 - val_loss: 0.1644 - val_binary_accuracy: 0.9455
Epoch 198/200
30000/30000 [==============================] - 2s 50us/sample - loss: 0.1432 - binary_accuracy: 0.9497 - val_loss: 0.1657 - val_binary_accuracy: 0.9462
Epoch 199/200
30000/30000 [==============================] - 2s 50us/sample - loss: 0.1428 - binary_accuracy: 0.9496 - val_loss: 0.1655 - val_binary_accuracy: 0.9469
Epoch 200/200
30000/30000 [==============================] - 2s 50us/sample - loss: 0.1422 - binary_accuracy: 0.9499 - val_loss: 0.1630 - val_binary_accuracy: 0.9465
```

图 4-21　执行结果

> **说　明**
>
> 前面介绍的二分类交叉熵其实是内建的 Sigmoid 激活函数，不过大多数人都习惯在网络层最后加上 Sigmoid 激活函数。为了避免输出执行两次 Sigmoid 的运算，keras.losses.BinaryCrossentropy 损失函数会有 from_logits 参数：
> （1）from_logits 为 False（默认）：损失函数不会加上 Sigmoid 激活函数运算。
> （2）from_logits 为 Ture：损失函数会加上 Sigmoid 激活函数运算。
> 第 5 章 keras.losses.CategoricalCrossentropy 损失函数是相同的概念。

Step 09 使用独热编码训练网络（Model 2）。

- 建立网络模型

网络架构与 4.3.3 小节"使用数值编码训练网络"中的网络架构相同。网络模型的输入为两只宝可梦的数据，每只宝可梦有 27 种不同的数据，所以两只宝可梦输入大小为(54,)。

```python
inputs = keras.Input(shape=(54, ))
x = layers.Dense(64, activation='relu')(inputs)
x = layers.Dropout(0.3)(x)
x = layers.Dense(64, activation='relu')(x)
x = layers.Dropout(0.3)(x)
x = layers.Dense(64, activation='relu')(x)
x = layers.Dropout(0.3)(x)
x = layers.Dense(16, activation='relu')(x)
x = layers.Dropout(0.3)(x)
outputs = layers.Dense(1, activation='sigmoid')(x)

model_2 = keras.Model(inputs, outputs, name='model-2')
# 显示网络架构
model_2.summary()
```

结果如图 4-22 所示。

```
Model: "model-2"
_____
Layer (type)                 Output Shape              Param #
=================================================================
input_2 (InputLayer)         [(None, 54)]              0
_____
dense_5 (Dense)              (None, 64)                3520
_____
dropout_4 (Dropout)          (None, 64)                0
_____
dense_6 (Dense)              (None, 64)                4160
_____
dropout_5 (Dropout)          (None, 64)                0
_____
dense_7 (Dense)              (None, 64)                4160
_____
dropout_6 (Dropout)          (None, 64)                0
_____
dense_8 (Dense)              (None, 16)                1040
_____
dropout_7 (Dropout)          (None, 16)                0
_____
dense_9 (Dense)              (None, 1)                 17
=================================================================
Total params: 12,897
Trainable params: 12,897
Non-trainable params: 0
```

图 4-22 执行结果

- 设置训练使用的优化器、损失函数和评价指标函数

```
model_2.compile(keras.optimizers.Adam(),
            loss=keras.losses.BinaryCrossentropy(),
            metrics=[keras.metrics.BinaryAccuracy()])
```

- 设置回调函数

```
# 将训练记录存成 TensorBoard 的记录文件
log_dir = os.path.join('lab3-logs', 'model-2')
model_cbk = keras.callbacks.TensorBoard(log_dir=log_dir)
# 存储最好的网络模型权重
model_mckp = keras.callbacks.ModelCheckpoint(model_dir + '/Best-model-2.h5',
                                    monitor='val_binary_accuracy',
                                    save_best_only=True,
                                    mode='max')
```

- 训练网络模型

```
history_2 = model_2.fit(x_train_one_hot, y_train,
            batch_size=64,
            epochs=200,
            validation_data=(x_val_one_hot, y_val),
            callbacks=[model_cbk, model_mckp])
```

结果如图 4-23 所示。

```
Epoch 195/200
30000/30000 [==============================] - 2s 51us/sample - loss: 0.0688 - binary_accuracy: 0.9715 - val_loss: 0.0980
 - val_binary_accuracy: 0.9652
Epoch 196/200
30000/30000 [==============================] - 2s 51us/sample - loss: 0.0654 - binary_accuracy: 0.9741 - val_loss: 0.1056
 - val_binary_accuracy: 0.9623
Epoch 197/200
30000/30000 [==============================] - 2s 51us/sample - loss: 0.0687 - binary_accuracy: 0.9725 - val_loss: 0.0997
 - val_binary_accuracy: 0.9637
Epoch 198/200
30000/30000 [==============================] - 2s 51us/sample - loss: 0.0686 - binary_accuracy: 0.9737 - val_loss: 0.1024
 - val_binary_accuracy: 0.9638
Epoch 199/200
30000/30000 [==============================] - 2s 51us/sample - loss: 0.0687 - binary_accuracy: 0.9722 - val_loss: 0.1010
 - val_binary_accuracy: 0.9632
Epoch 200/200
30000/30000 [==============================] - 2s 51us/sample - loss: 0.0690 - binary_accuracy: 0.9725 - val_loss: 0.1027
 - val_binary_accuracy: 0.9629
```

图 4-23　执行结果

Step 10 比较两种网络的训练结果。

- 显示两个网络的历史正确率

从结果图可以发现 model-2 的训练和验证都高于 model-1，证实了独热编码的作用。

```
plt.plot(history_1.history['binary_accuracy'], label='model-1-training')
plt.plot(history_1.history['val_binary_accuracy'], label='model-1-validation')
plt.plot(history_2.history['binary_accuracy'], label='model-2-training')
plt.plot(history_2.history['val_binary_accuracy'], label='model-2-validation')
plt.ylabel('Accuracy')
plt.xlabel('epochs')
plt.legend()
```

结果如图 4-24 所示。

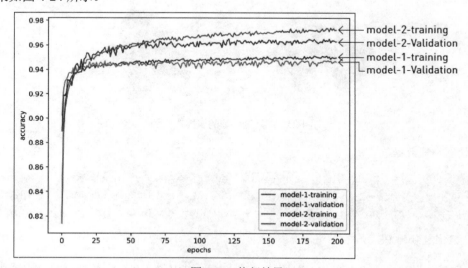

图 4-24　执行结果

- 在测试集数据上验证

测试结果表明 model-2 优于 model-1。

```python
# 加载 Model 1 正确率最高的模型权重
model_1.load_weights(model_dir + '/Best-model-1.h5')
# 加载 Model 2 正确率最高的模型权重
model_2.load_weights(model_dir + '/Best-model-2.h5')
loss_1, accuracy_1 = model_1.evaluate(x_test_normal, y_test)
loss_2, accuracy_2 = model_2.evaluate(x_test_one_hot, y_test)
print("Model-1: {}%\nModel-2: {}%".format(accuracy_1, accuracy_2))
```

结果如下：

```
10000/10000 [==============================] - 0s 36us/sample
            - loss: 0.1593 - binary_accuracy: 0.9466
10000/10000 [==============================] - 0s 36us/sample
            - loss: 0.0947 - binary_accuracy: 0.9654
Model-1: 0.9466000199317932%
Model-2: 0.965399980545044%
```

Step 11 宝可梦大 PK。

最后我们进行一个实验，使用最初拿到的 3 只宝可梦的最终进化进行 PK，分别为妙蛙花、喷火龙和水箭龟，如图 4-25 所示，分别对应编号 3、7 和 12。

图 4-25　3 只宝可梦的属性克制图

- 读取个别数据

```
venusaur = np.expand_dims(pokemon_data_one_hot[3], axis=0)    # 妙蛙花
charizard = np.expand_dims(pokemon_data_one_hot[7], axis=0)   # 喷火龙
blastoise = np.expand_dims(pokemon_data_one_hot[12], axis=0)  # 水箭龟
```

- 3 只宝可梦的 PK 预测

最终的预测结果喷火龙胜过了妙蛙花与水箭龟，结果出乎意料，因为这个预测并没有按照常理的属性克制来预测，而是跟宝可梦动画的剧情一样，喷火龙战胜水箭龟。

```
# 妙蛙花 vs 喷火龙
pred = model_2.predict(np.concatenate([venusaur, charizard], axis=-1))
```

```
winner = '妙蛙花' if pred < 0.5 else '喷火龙'
print("pred={}, {} 获胜".format(pred, winner))

# 喷火龙 vs 水箭龟
pred = model_2.predict(np.concatenate([charizard, blastoise], axis=-1))
winner = '喷火龙' if pred < 0.5 else '水箭龟'
print("pred={}, {} 获胜".format(pred, winner))

# 水箭龟 vs 妙蛙花
pred = model_2.predict(np.concatenate([blastoise, venusaur], axis=-1))
winner = '水箭龟' if pred < 0.5 else '妙蛙花'
print("pred={}, {} 获胜".format(pred, winner))
```

结果如下：

```
pred=[[1.]], 喷火龙获胜
pred=[[1.0699459e-07]], 喷火龙获胜
pred=[[0.9999981]], 妙蛙花获胜
```

4.4 参考文献

[1] Keller J M, Gray M R, Givens J A. A fuzzy k-nearest neighbor algorithm[J]. In IEEE transactions on systems, 1985, 580-585.

[2] Denoeux T. A k-nearest neighbor classification rule based on Dempster-Shafer theory[J]. in IEEE Transactions on Systems, Man, and Cybernetics, 1995, 25(5):804-813.

[3] Fukunaga K, Narendra P M. A Branch and Bound Algorithm for Computing k-Nearest Neighbors. in IEEE Transactions on Computers, 1975, C-24(7):750-753.

[4] Dudani S A. The Distance-Weighted k-Nearest-Neighbor Rule[J]. In IEEE Transactions on Systems, Man, and Cybernetics, 1976, SMC-6(4):325-327.

[5] Quinlan J R. Induction of decision trees[J]. Machine Learning, 1986, 81-106.

[6] Schmid H. Probabilistic part of speech tagging using decision trees[C]. In Proceedings of the International Conference on New Methods in Language Processing, 1994, 44-49.

[7] Janikow C Z. Fuzzy decision trees: issues and methods[J]. in IEEE Transactions on Systems, Man, and Cybernetics, Part B (Cybernetics), 1998, 28(1):1-14.

[8] Tsang S, Kao B, Yip K Y, et al. Decision Trees for Uncertain Data[J]. in IEEE Transactions on Knowledge and Data Engineering, 2011, 23(1): 64-78.

[9] Breiman L. Random Forests[J]. Machine Learning, 2001, 45(1):5-32.

[10] Bosch A, Zisserman A, Munoz X. Image Classification using Random Forests and Ferns[C]. IEEE International Conference on Computer Vision, 2007, 1-8.

[11] Gislason P O, Benediktsson J A, Sveinsson J R. Random forests for land cover classification[J]. Pattern Recognit. Lett, 2006, 27(4):294-300.

[12] Cortes C, Vapnik V. Support-Vector Network[J]. Machine Learning, 1995, 273-297.

[13] Joachims T. Text categorization with support vector machines: Learning with many relevant features[C]. In European Conference on Machine Learning, 1998, 137-142.

[14] Campbell W M, Sturim D E, Reynolds D A. Support vector machines using GMM supervectors for speaker verification[J]. In IEEE Signal Processing Letters, 2006, 13(5):308-311.

[15] Bennett K P, Demiriz A. Semi-supervised support vector machines[C]. In Advances in Neural Information processing systems, 1999, 368-374.

[16] LeCun Y, Bengio Y, Hinton G. Deep learning[J]. Nature, 2015, 521(7553):436-444.

[17] Krizhevsky A, Sutskever I, Hinton G E. Imagenet classification with deep convolutional neural networks[C]. Advances in Neural Information Processing Systems, 2012, 1097-1105.

[18] Le T, Huang S, Jaw D. Cross-Resolution Feature Fusion for Fast Hand Detection in Intelligent Homecare Systems[J]. in IEEE Sensors Journal, 2019, 19(12):4696-4704.

[19] Liu Y, Jaw D, Huang S, et al. DesnowNet: Context-Aware Deep Network for Snow Removal[J]. in IEEE Transactions on Image Processing, 2018, 27(6):3064-3073.

[20] Hartigan J A, Wong M A. A K-Means Clustering Algorithm[J]. Applied Statistics, 1979, 28(1):100-108.

[21] Jain A K. Data clustering: 50 years beyond K-means[J]. Pattern Recognit. Lett., 2010, 31(8):651-666.

[22] Kanungo T et al. An efficient k-means clustering algorithm: Analysis and implementation[J]. IEEE Transactions on Pattern Analysis and Machine Intelligence, 2002, 24(7):881-892.

[23] Selim S Z, Ismail M A. K-Means-Type Algorithms: A Generalized Convergence Theorem and Characterization of Local Optimality[J]. in IEEE Transactions on Pattern Analysis and Machine Intelligence, 1984, PAMI-6(1):81-87.

[24] Navarro J F, Frenk C S, White S D M. A universal density profile from hierarchical clustering[J]. Astrophysical J, 1997, 490(2):493-508.

[25] Bajcsy P, Ahuja N. Location-and density-based hierarchical clustering using similarity analysis[J]. in IEEE Transactions on Pattern Analysis and Machine Intelligence, 1998, 20(9):1011-1015.

[26] Tang X, Zhu P. Hierarchical Clustering Problems and Analysis of Fuzzy Proximity Relation on Granular Space[J]. in IEEE Transactions on Fuzzy Systems, 2013, 21(5):814-824.

[27] Wold S, Esbensen K, Geladi P. Principal component analysis[J]. Chemometr. Intell. Lab. Syst, 1987, 37-52.

[28] Joliffe I. Principal Component Analysis[D]. New York: Springer-Verlag, 1986.

[29] Moore B. Principal component analysis in linear systems: Controllability, observability, and model reduction[J]. in IEEE Transactions on Automatic Control, 1981, 26(1):17-32.

[30] Tipping M E. Sparse kernel principal component analysis[C]. in Neural Information Processing Systems, 2000, 633-639.

[31] Mika S, Scholkopf B, Smola A, et al. Kernel PCA and de-noising in feature spaces[C]. Advances in Neural Information Processing Systems, 1999, 11(1):536-542.

[32] Chengjun Liu. Gabor-based kernel PCA with fractional power polynomial models for face recognition[J]. in IEEE Transactions on Pattern Analysis and Machine Intelligence, 2004, 26(5):572-581.

[33] Hinton G, Salakhutdinov R. Reducing the dimensionality of data with neural networks[J]. Science, 2006, 313(5786):504-507.

[34] Wang W, Huang Y, Wang Y, et al. Generalized autoencoder: A neural network framework for dimensionality reduction[C]. Proceedings of the IEEE Conference on Computer Vision and Pattern Recognition Workshops, 2014, 490-497.

[35] Silver D. Mastering the game of Go with deep neural networks and tree search[J]. Nature, 2016, 529(7587):484-489.

[36] Silver D. Mastering the game of go without human knowledge[J]. Nature, 2017, 550(7676):354-359.

第 5 章

多分类问题

学习目标

- 掌握卷积神经网络的基本概念及原理
- 了解 Softmax 激活函数的使用
- 了解多分类交叉熵的使用
- 使用数据增强方法增加数据量
- 运用卷积神经网络完成 CIFAR-10 分类任务

5.1 卷积神经网络

5.1.1 卷积神经网络简介

卷积神经网络（Convolutional Neural Network，CNN）近年来在图像识别和声音识别方面都有不错的效果，特别是图像识别，目前几乎所有的视觉竞赛中都是以卷积神经网络为基础的。在第 3 章介绍过的 ImageNet 大赛冠军 AlexNet[1]网络也是以卷积神经网络为基础搭建的，图 5-1 所示为 2012 年 ImageNet[2]大赛的排行榜，使用卷积神经网络的 AlexNet 遥遥领先，而之后每一年的 ImageNet 大赛都是使用卷积神经网络架构的网络模型获得第一。

卷积神经网络主要是对图像进行局部的特征提取，每一层卷积负责检测的特征都不同，像是前面的卷积层检测低级的特征（线条、颜色、纹路等），而后面的卷积层会检测高级的特征（眼睛、鼻子、手等），在卷积神经网络特征提取后，通常会通过全连接层（Fully Connected Layer）进行分类，例如人的图像会分类到人的类别，狗的图像会分类到狗的类别。

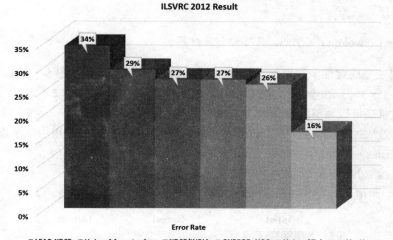

图 5-1　ImageNet 2012 的排行榜

说　明
为什么不使用全连接架构处理图像呢？主要有两种原因： 1. 全连接层输入维度：全连接层要求一维张量（Tensor）的输入，所以输入图像会被重塑（Reshape）成一维的张量，进而丢失图像空间信息。 2. 全连接的处理方式：全连接层是对输入的所有像素进行分类，所以全连接层考虑的是全局的信息，但实际上我们在观察物体时并不会观察背景等无关要紧的全局信息，而是观察物体本身（局部信息）。

5.1.2　卷积神经网络架构

一个典型的分类卷积神经网络如图 5-2 所示，由卷积层（Convolution Layer）、池化层（Pooling Layer）、压平层（Flatten Layer）和全连接层（Fully Connected Layer）组成。下面将会按序介绍卷积层、池化层和压平层。

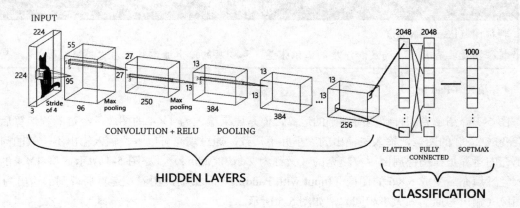

图 5-2　卷积神经网络

卷积层

- 卷积核（Convolution Kernel，又可称作过滤器）

卷积层的原理是通过许多卷积核（简称为 Kernel）在图像上滑动，图 5-3 所示为输入（Input）4×4 的图像和一个 3×3 的卷积核，卷积核在图像上滑动，最终产生 2×2 的输出（Output）。卷积核的大小和数量是可以调整的超参数，大多数大小都设置为奇数（3×3、5×5 或 7×7）。卷积核的数量太多会造成过拟合问题，需经过调参挑选出适合的参数，并且卷积层同全连接层可以选择是否加上偏差（Bias）参数。

图 5-3 卷积运算的第一行输出为 7，计算公式如下：

(1×1) + (1×1) + (0×0) + (2×0) + (3×−1) + (2×2) + (1×2) + (4×0) + (2×1) = 7

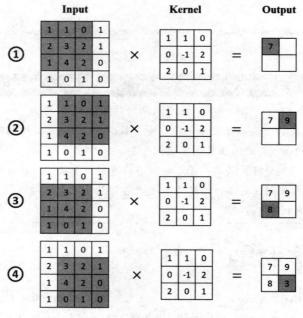

图 5-3　卷积运算

说　明
特征图（Feature Map） 特征图是具有长度、高度和深度的三维张量，卷积层的输出通常被称为"特征图"。

- 填充（Padding）

图 5-3 介绍了卷积神经网络基础的运算，但是原来输入为 4×4 的图像，经过卷积运算后，输出会变成 2×2 的图像，输入与输出的大小并不一致，通过填充可以解决输入输出不一致的问题。而填充的运算是在图像周围用零填充的（又称为 Zero-Padding），如图 5-4 所示，将 4×4 的图像输入经过填充后变成 6×6 的图像（Input with Padding），最后经由卷积运算后得到的输出为 4×4 的图像（输出与原输入大小相同），如图 5-5 所示。

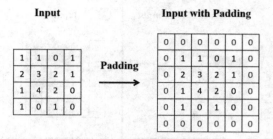

图 5-4 填充运算（输入 4×4 的图像，经过填充得到 6×6 的图像）

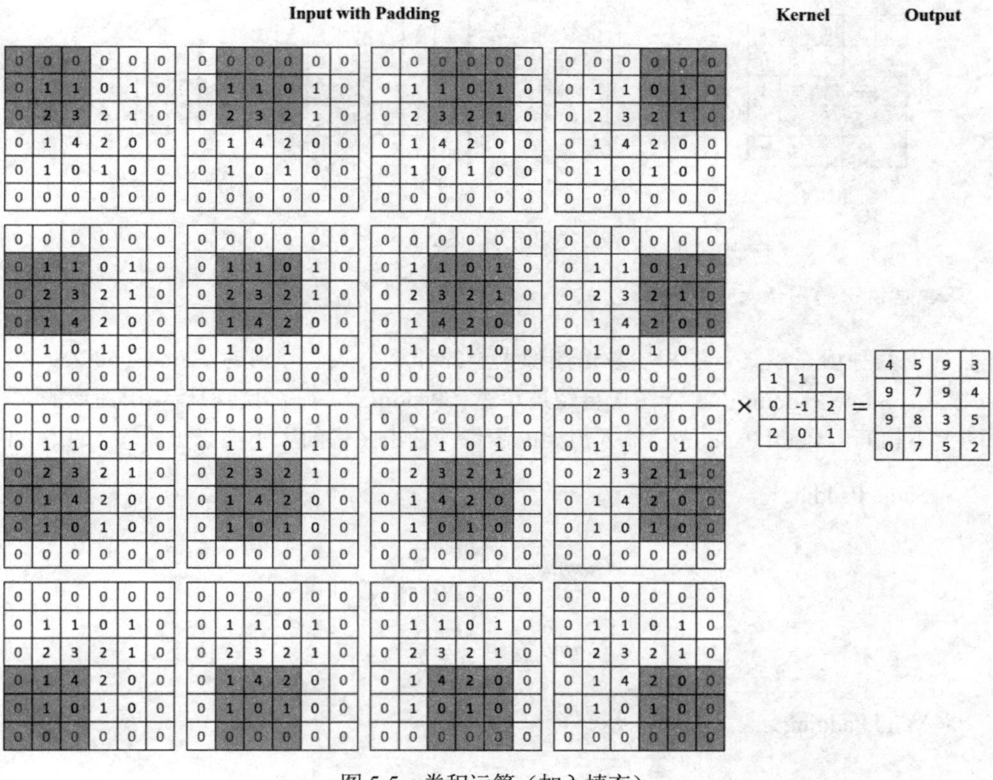

图 5-5 卷积运算（加入填充）

- 步长（Stride）

一般卷积层运算使用的步长都为 1，也可以将其设置为大于 1 的值，这种卷积又称为"步长卷积"（Strided Convolution）。例如，输入图像为 5×5 的大小，卷积运算步长设置为 2（代表卷积核每一次会滑动两步），最终为 2×2 的输出，如图 5-6 所示。Stride 为 2 的卷积层对输入进行了两倍的下采样（Subsampled，缩小图像），得到比原输入还小的图像，使得后方卷积的运算量减少。

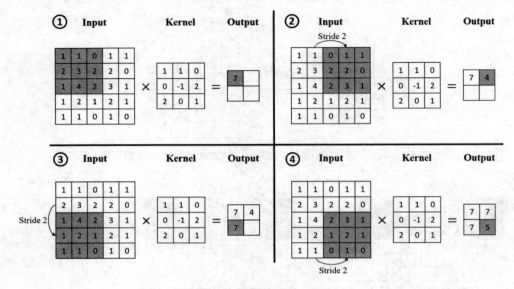

图 5-6　卷积运算（Stride=2）

- 卷积层输出的图像大小的计算

下面列出两种填充模式下卷积层输出图像大小的计算公式，分别为 Same Padding 和 Valid Padding 模式。Same Padding：输入图像会经过填充（Padding），使得卷积后输出图像的大小与输入图像的大小相同。Valid Padding：输入原始的图像，输出图像的大小会小于原输入图像的大小。

➢ Same Padding：

$$Output_{height} = \frac{Input_{height}}{Strides}$$

$$Output_{width} = \frac{Input_{width}}{Strides}$$

➢ Valid Padding：

$$Output_{height} = \frac{Input_{height} - kernel_{height} + 1}{Strides}$$

$$Output_{width} = \frac{Input_{width} - kernel_{width} + 1}{Strides}$$

Input：输入图像的大小。
Output：输出图像的大小。
kernel：卷积核的大小。
Strides：卷积运算的步长。

前面的范例都是单个输入 Channel 和单个 Kernel，但实际上卷积运算都是多个输入 Channel 和多个 Kernel，如图 5-7 所示，对输入的彩色图像进行卷积运算。

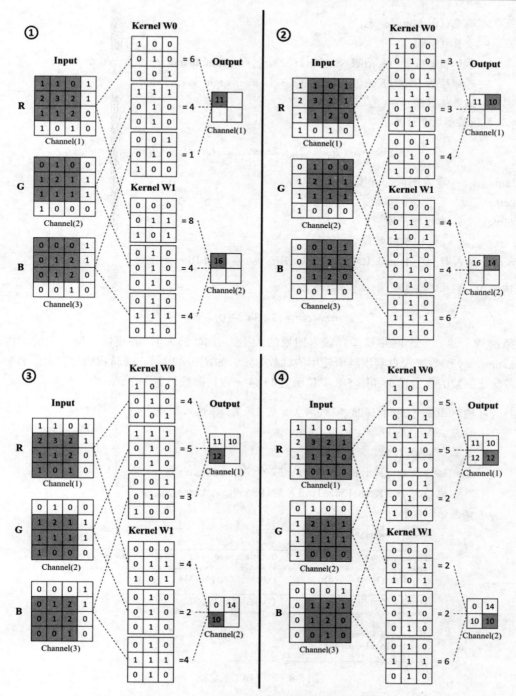

图 5-7 对输入为 4×4×3 的图像进行卷积运算（卷积核数量为 2）

参数设置如下：

> 输入图像（Input）：4×4×3（长度，宽度，深度），图像的输入深度（Input_channel）为 R（Red）、G（Green）、B（Blue）3 种颜色。

- 填充（Padding）：无。
- 步长（Stride）：1。
- 卷积核的数量（Kernel Number）：2 个（kernelnumbers，分别为 W0 和 W1）。
- 卷积核的大小（Kernel Size）：3×3（kernelheight, kernelwidth）。
- 偏差值（Bias）：无。

● 卷积层参数量计算

$$Parameter = (Input_{channel} \times kernel_{height} \times kernel_{width} + Bias) \times kernel_{numbers}$$

$Input_{channel}$：输入图像的深度。
$kernel_{height}$：卷积核的高度。
$kernel_{width}$：卷积核的宽度。
$kernel_{numbers}$：卷积核的数量。
Bias：是否有加上 Bias，如果有，那么 Bias = 1，否则 Bias = 0。

范例 1 图 5-7 的卷积层参数计算。

$$Parameter = (3 \times 3 \times 3 + 0) \times 2 = 54$$

范例 2 建立一个网络模型，输入图像的大小为 28×28×4，并连接一层卷积层，再通过 model.summary 函数来观察卷积层所使用的参数数量，如图 5-8 所示。最后我们使用上面的参数计算公式验证参数的数量是否和图 5-8 中显示的 1184 一致，参数设置如下：

- 输入图像（Input）：28×28×4（长度, 宽度, 深度）。
- 填充（Padding）：无。
- 步长（Stride）：1。
- 卷积核的数量（Kernel Number）：32 个（kernelnumbers）。
- 卷积核的大小（Kernel Size）：3×3（kernelheight, kernelwidth）。
- 偏差值（Bias）：有。

```
Model: "model"
_____
Layer (type)                 Output Shape              Param #
=================================================================
input_1 (InputLayer)         [(None, 28, 28, 4)]       0
_____
conv2d (Conv2D)              (None, 26, 26, 32)        1184
=================================================================
Total params: 1,184
Trainable params: 1,184
Non-trainable params: 0
_____
```

图 5-8　网络模型示意图

程序代码如下：

```
from tensorflow import keras
inputs = keras.Input((28, 28, 4))
outputs = keras.layers.Conv2D(32, kernel_size=3, strides=(1, 1), padding='valid', use_bias=True)(inputs)
```

```
model = keras.Model(inputs, outputs)
model.summary()
```

卷积层参数计算：计算结果与图 5-8 所示的参数数量一致。

$$Parameter = (4 \times 3 \times 3 + 1) \times 32 = 1184$$

 池化层

池化层常用来进行下采样的运算，与卷积层不同，池化层并没有参数，只有 Max 或 Mean 的张量运算。池化层有许多种方法，例如最大池化（Max Pooling）或平均池化（Average Pooling）等。图 5-9 所示为最大池化层的运算，以 4×4 图像作为输入（Input），最大池化层会去计算 2×2 区块的最大值作为输出，并且每计算完一个区块会向右或向下滑动两格，最终产生 2×2 的输出。一般来说，平均池化层保留更多的背景信息，而最大池化层则保留更多的纹理信息。

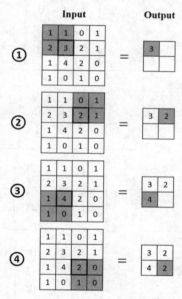

图 5-9　最大池化层的运算

 压平层

压平层大多用于卷积层与全连接层之间，作为两层之间的桥梁，负责将卷积层提取到的多维特征张量压平成一维特征张量，并传入全连接层进行分类。例如，图 5-10 把输入为 4×4 的二维张量图像压平成 16 维的向量（一维张量）。

图 5-10　压平层运算

5.1.3 卷积神经网络的原理

■ 卷积核的作用

卷积层中具有多个不同的卷积核（又称为过滤器），通过不同过滤器在图像上滑动可以产生不同作用，例如边缘检测、模糊、锐化或检测线条等功能，如表 5-1 所示。但这些过滤器并非刚开始就具有检测功能，起初过滤器都不具有任何意义，需经过网络学习更新参数，进而学习到对识别有用的过滤器。

表 5-1 不同的过滤器参数所代表的意义[3]

图像输入	运算	Kernel 参数	图像输出
		$\begin{bmatrix} 0 & 0 & 0 \\ 0 & 1 & 0 \\ 0 & 0 & 0 \end{bmatrix}$	
	边缘检测	$\begin{bmatrix} 1 & 0 & -1 \\ 0 & 0 & 0 \\ -1 & 0 & 1 \end{bmatrix}$	
		$\begin{bmatrix} -1 & -1 & -1 \\ -1 & 8 & -1 \\ -1 & -1 & -1 \end{bmatrix}$	
	锐化	$\begin{bmatrix} 0 & -1 & 0 \\ -1 & 5 & -1 \\ 0 & -1 & 0 \end{bmatrix}$	
	高斯模糊	$\begin{bmatrix} 1 & 2 & 1 \\ 2 & 4 & 2 \\ 1 & 2 & 1 \end{bmatrix}$	

不过，上面所说明的都只有一层卷积层，如果将网络层加深后，就很难想象输出会变成怎样，不过也不是很难解释。我们可以将深层的卷积网络想象为图 5-11，前面几层的卷积层（1~3）主要负责识别边缘或线条等，后面几层的卷积层（4~6）基于前面提取到的特征，能够识别更为具体的特征，例如鼻子、眼睛或耳朵等，最后通过压平层将输出特征图压平成一维张量，再经过全连接层将卷积层提取到的特征进行分类。

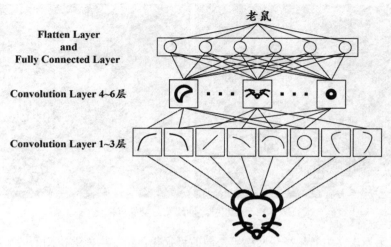

图 5-11　深层卷积网络理解

可视化卷积神经网络

前面已经讲述了卷积神经网络前面几层负责识别边缘或线条等简单的特征，后面几层则负责识别更加具体的特征，例如鼻子或眼睛。目前已经有许多方法可以观察卷积层学习到的特征[4]-[7]，这里介绍的 Visualizing and Understanding Convolutional Network[7]，经由论文就可以验证卷积神经网络的这个特性。首先，卷积神经网络第一层的可视化如图 5-12 所示，图中网格的每张图像都代表一种识别模式，如左上角图像为识别-45 度的线、中间顶部图像为识别+45 度的线，而图 5-13 的九宫格中左上角九张图像激活了-45 度线的输入，中间顶部的九张图像则是激活了+45 度线的输入。

图 5-12　卷积神经网络第一层的可视化[1]

图 5-13　卷积神经网络第一层的可视化[1]

图 5-14 所示为论文中卷积神经网络第二层的可视化，第二层能够识别圆圈和条纹等特征，如图中第二行、第二列识别圆圈，以及第一行、第二列识别条纹。

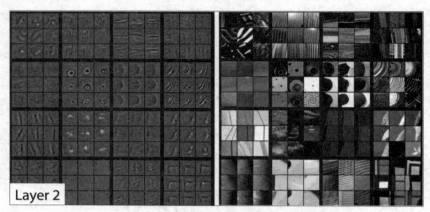

图 5-14 卷积神经网络第二层的可视化

图 5-15 所示为论文中卷积神经网络第三层的可视化,第三层能够检测的图形更加复杂,如图中的左上方能识别蜂窝,第二行、第二列识别轮子。

图 5-15 卷积神经网络第三层的可视化

图 5-16 所示为论文中卷积神经网络第四层和第五层的可视化,也是整个架构的最后两层卷积层,这两层可以识别的东西变得十分具体,例如可以识别鸟的脚、狗的头、动物的眼睛、键盘、脚踏车,甚至接近个体了。

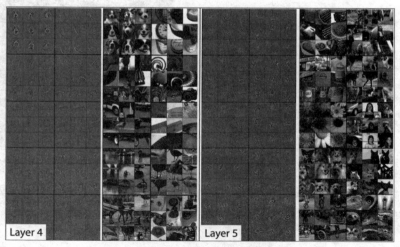

图 5-16 卷积神经网络第四层和第五层的可视化

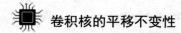
卷积核的平移不变性

不变性（Invariance）代表图像经过平移、翻转、缩放甚至不同的光线条件都可以成功被识别出来。这里要讨论的平移不变性指的是图像中的目标无论出现在左上角、右下角或图像中的任何地方，得到的结果应该是相同的。

下面列出了几种不变性：

- 平移不变性：目标在不同位置能够得到相同的结果。
- 缩放不变性：目标放大或缩小能够得到相同的结果。
- 旋转不变性：目标旋转方向能够得到相同的结果。
- 光照不变性：目标颜色、亮度或饱和度改变能够得到相同的结果。

对于全连接网络，如果目标出现在不同的位置，全连接层的输出就会全然不同，所以只能重新学习不同位置的数据来克服；而对于卷积神经网络，如果将图像向右移动，卷积产生的输出结果就是一样的，这是因为卷积层的过滤器是在图像上平移运算产生输出，所以可以用下面这个等式表示卷积层的平移不变性。

$$F(T(x)) = T(F(x))$$

x：输入图像。
F：卷积层的运算。
T：平移的转换。

5.2 多分类问题

5.2.1 Softmax

在设计多分类问题（Multiclass Classification Problem）的深度网络时，最后一层网络架构通常使用 Softmax 激活函数（又称为归一化指数函数），主要是将神经网络预测的输出结果精确地用概率模型来描述，经过 Softmax 激活函数后，每个输出值都介于 0 和 1 之间，且保证所有输出值的总和为 1。计算公式如下：

$$y_i = \frac{e^{z_i}}{\sum_{j=0}^{C} e^{z_j}}$$

C：类别总数。
y：预期输出。
z：深度学习模型的预测值。
i：深度学习模型第 i 个输出。

以图 5-17 为例，网络要判断输入数据属于 3 个类别中的哪一类，输入 $x1$、$x2$、$x3$，经过网络模型运算得到输出结果 3、1、-3，虽然可以知道结果属于第一类，但是经过 Softmax 函数运算后，将 3、1、-3 转换成 0.88、0.12、0，可以明确地表示属于第一类的可能性最大。

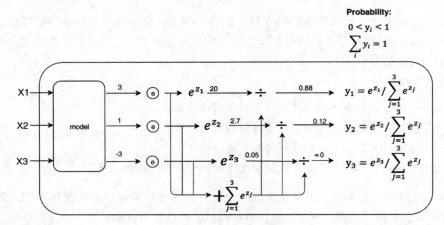

图 5-17 Softmax 输出运算

5.2.2 多分类交叉熵

回顾 4.2.3 小节所介绍的分类问题，其中使用的损失函数为交叉熵（Cross-Entropy，CE），公式如下：

$$CE = -\frac{\sum_{i=1}^{N}\sum_{j=0}^{C} y_{i,j} \log \hat{y}_{i,j}}{N}$$

y：预期输出值。
$\hat{y}$：深度学习模型的预测输出值（预测输出值通常会先经过 Sigmoid 或 Softmax 激活函数）。
C：类别数量。
N：一个批量的数据量。

第 4 章已介绍过分类问题又可以分为二分类问题和多分类问题，二分类问题通常使用二分类交叉熵作为损失函数，多分类问题则使用多分类交叉熵作为损失函数。

本小节的范例为多分类问题，所以使用多分类交叉熵作为损失函数，多分类交叉熵损失函数为 Softmax 激活函数和交叉熵损失函数的组合，所以又称为 Softmax Loss，如图 5-18 所示。

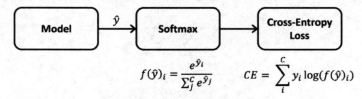

图 5-18 多分类交叉熵

多分类交叉熵公式如下：

$$CCE = -\frac{\sum_{i=1}^{N}\sum_{j=0}^{C} y_{i,j} \log(f(\hat{y}_{i,j}))}{N}$$

y：预期输出值。
$\hat{y}$：深度学习模型经过 Softmax 后的预测值。
f：Softmax 函数。
C：类别数量。
N：一个批量的数据量。

多分类交叉熵的计算以图 5-19 为例，输入猫的图像，经过神经网络得到预测结果 [0,0,0,0.6,0,0,0.1,0,0.3,0]，与真实值（y_true）[0,0,0,1,0,0,0,0,0,0]之间的误差计算如下：

$$\begin{aligned} loss &= -(0 \times \log 0 + 0 \times \log 0 + 0 \times \log 0 + 1 \times \log 0.6 + 0 \times \log 0 \\ &\quad + 0 \times \log 0 + 0 \times \log 0.1 + 0 \times \log 0 + 0 \times \log 0.3 + 0 \times \log 0) \\ &= -1 \times \log 0.6 \\ &= 0.22 \end{aligned}$$

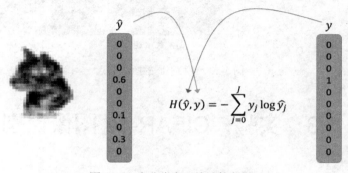

图 5-19　多分类交叉熵计算范例

5.2.3　数据增强

第 3 章已经介绍过深度学习经常会遇到过拟合的问题，而较佳的解决方法是加入更多的训练数据来避免过拟合发生。本小节会介绍数据增强（Data Augmentation）方法，此方法是通过计算机运算转换，而非手动收集数据产生的，所以能大量减少收集数据的时间[8]-[13]。这项技术常应用于图像处理的问题上，所以又称为图像增强（Image Augmentation），通过简单的图像翻转、图像缩放、图像旋转或颜色转换就可以产生原来数据两倍以上的数据量，如图 5-20 所示。下面列出了常见的图像增强方法。

- 图像翻转。
- 图像旋转。
- 图像平移。
- 图像缩放。
- 颜色转换（对比度、饱和度或亮度等）。

- 图像模糊（高斯模糊或平均模糊等）。
- 加入噪声（高斯噪声或胡椒盐噪声等）。
- 加入气候环境（雾、雨或雪等）。

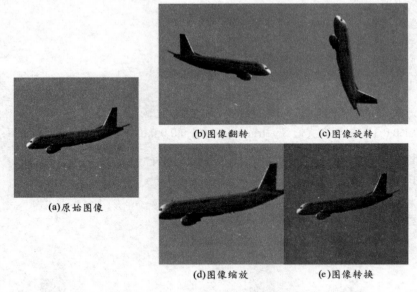

图 5-20　4 种不同的图像增强

5.3　实验：CIFAR-10 图像识别

本节主题为 CIFAR-10 图像识别，希望通过神经网络模型来预测出图像的类别。网络模型训练使用多元交叉熵作为损失函数，Adam 作为模型优化器，并训练 3 个网络模型，比较它们的性能差异，网络模型的差异分别如下：

- Model-1：主要使用全连接层搭建网络模型，模型的参数数量会尽量设计得与 Model-2 和 Model-3 一样。
- Model-2：主要使用卷积层搭建网络模型。
- Model-3：网络架构与 Model-2 完全一样，但会加入图像增强来增加数据的多样性。

最后程序的实验结果显示模型之间的性能为 Model-3→Model-2→Model-1。

5.3.1　数据集介绍

本小节使用的数据集为 CIFAR-10，数据集包含 60 000 张 32×32 的 RGB 图像，共分为 10 个类别（Airplane、Automobile、bird、cat、deer、dog、frog、horse、ship、truck），如图 5-21 所示，其中 50 000 张为训练数据，10 000 张为测试数据。读者可以自行到官方网站（https://www.cs.toronto.edu/~kriz/cifar.html）下载数据集。下面将介绍一个更加方便快捷的方法。

图 5-21　CIFAR-10 数据集

5.3.2　TensorFlow Datasets

　　数据集的收集在深度学习中扮演着非常重要的角色，但每个数据集的存储格式和表达方式都不尽相同。例如，同样为分类数据集，存储标签的文件有可能为 TXT 文件、XML 文件或 JSON 文件等。由于有太多种存储方式和格式，导致每次编写读取和处理数据的脚本都需要花费很多时间，而且还有错误的风险，所以这里介绍 TensorFlow Datasets 就是为了改善上述问题。

　　TensorFlow Datasets 是一个拥有多个开放数据集的数据库，每个数据集都被整理成 **tf.data.dataset** 的形式，通过几行指令即可完成数据的读取，并可以使用 tf.data API 搭建高性能的输入管道，例如下面简短的范例程序。

```python
import tensorflow_datasets as tfds

# 加载数据集
train_data = tfds.load("mnist", split= tfds.Split.TRAIN)
# 设置输入管道
train_data = train_data.shuffle(1024).batch(32)
            .prefetch(tf.data.experimental.AUTOTUNE)
# 训练网络
model.fit(train_data, epochs=100)
```

　　还有一点值得注意的是，TensorFlow Datasets 这个平台是开放的，所以任何人都可以在平台上建立公开的数据集，官方文件也提供了范例程序代码，希望能够收集到更多的公开数据集，让更多研究人员、工程师和使用者受益。目前，TensorFlow Datasets 上的数据集已超过 52 个以上。

　　TensorFlow Datasets 官方网址为 https://www.tensorflow.org/datasets/。

5.3.3 新建项目

建议使用 Jupyter Notebook 来执行本小节的程序代码，操作流程如下：

Step 01 启动 Jupyter Notebook。

在 Terminal（Ubuntu）或命令提示符（Windows）中输入如下指令：

```
jupyter notebook
```

Step 02 新建执行文件。

单击界面右上角的 New 下拉按钮，然后单击所安装的 Python 解释器（在 Jupyter 中都称为 Kernel）来启动它，如图 5-22 所示，显示了 3 个不同的 Kernel：

- Python 3：本地端 Python。
- tf2：虚拟机 Python（TensorFlow-cpu 版本）。
- tf2-gpu：虚拟机 Python（TensorFlow-gpu 版本）。

图 5-22 新建执行文件

Step 03 执行程序代码。

按 Shift + Enter 快捷键执行单行程序代码，如图 5-23 所示。

图 5-23 Jupyter 环境界面

接下来，后续的程序代码都可以在 Jupyter Notebook 上执行。

5.3.4 程序代码

Step 01 导入必要的套件。

```
import os
import numpy as np
import pandas as pd
import tensorflow as tf
import matplotlib.pyplot as plt
from tensorflow import keras
from tensorflow.keras import layers
import tensorflow_datasets as tfds
```

Step 02 读取数据并分析。

- 查看 TensorFlow Datasets 目前提供的数据集

```
tfds.list_builders()
```

结果如下：

```
['abstract_reasoning',
 'bair_robot_pushing_small',
 'caltech101',
 'cats_vs_dogs',
 'celeb_a',
 'celeb_a_hq',
 'chexpert',
 'cifar10',
 'cifar100',
 'cifar10_corrupted',
 'cnn_dailymail',
 'coco2014',
 'colorectal_histology',
 'colorectal_histology_large',
 'cycle_gan',
 'diabetic_retinopathy_detection',
 'dsprites',
 'dtd',
 'dummy_dataset_shared_generator',
 'dummy_mnist',
 'emnist',
 'fashion_mnist',
 'flores',
 'glue',
 'groove',
 'higgs',
 'horses_or_humans',
 'image_label_folder',
 'imagenet2012',
```

```
 'imagenet2012_corrupted',
 'imdb_reviews',
 'iris',
 'kmnist',
 'lm1b',
 'lsun',
 'mnist',
 'moving_mnist',
 'multi_nli',
 'nsynth',
 'omniglot',
 'open_images_v4',
 'oxford_flowers102',
 'oxford_iiit_pet',
 'para_crawl',
 'quickdraw_bitmap',
 'rock_paper_scissors',
 'shapes3d',
 'smallnorb',
 'squad',
 'starcraft_video',
 'sun397',
 'svhn_cropped',
 'ted_hrlr_translate',
 'ted_multi_translate',
 'tf_flowers',
 'titanic',
 'ucf101',
 'voc2007',
 'wikipedia',
 'wmt15_translate',
 'wmt16_translate',
 'wmt17_translate',
 'wmt18_translate',
 'wmt19_translate',
 'wmt_translate',
 'xnli']
```

- 加载 CIFAR-10 数据集

```
# 将训练数据重新分成1:9等分，分别分给验证数据和训练数据
valid_split, train_split = tfds.Split.TRAIN.subsplit([10, 90])
# 获取训练数据，并读取数据的信息
train_data, info = tfds.load("cifar10", split=train_split, with_info=True)
# 获取验证数据
valid_data = tfds.load("cifar10", split=valid_split)
# 获取测试数据
test_data = tfds.load("cifar10", split=tfds.Split.TEST)
```

- 显示 CIFAR-10 数据集信息

可以从输出结果看到数据集的一些基本信息，例如输入图像的大小、类别数量、数据集划分方式等。

```
print(info)
```

结果如下：

```
tfds.core.DatasetInfo(
    name='cifar10',
    version=1.0.2,
    description='The CIFAR-10 dataset consists of 60000 32x32 colour images in 10 classes, with 6000 images per class. There are 50000 training images and 10000 test images.', urls=['https://www.cs.toronto.edu/~kriz/cifar.html'],
    features=FeaturesDict({
        'image': Image(shape=(32, 32, 3), dtype=tf.uint8),
        'label': ClassLabel(shape=(), dtype=tf.int64, num_classes=10)
    },
    total_num_examples=60000,
    splits={
        'test': <tfds.core.SplitInfo num_examples=10000>,
        'train': <tfds.core.SplitInfo num_examples=50000>
    },
    supervised_keys=('image', 'label'),
    citation='"""
        @TECHREPORT{Krizhevsky09learningmultiple,
            author = {Alex Krizhevsky},
            title = {Learning multiple layers of features from tiny images},
            institution = {},
            year = {2009}
        }
    """',
    redistribution_info=,
)
```

- 显示 CIFAR-10 的 10 个标签

```
labels_dict = dict(enumerate(info.features['label'].names))
labels_dict
```

结果如下：

```
{0: 'airplane',
 1: 'automobile',
 2: 'bird',
 3: 'cat',
 4: 'deer',
 5: 'dog',
 6: 'frog',
```

```
7: 'horse',
8: 'ship',
9: 'truck'}
```

- 查看训练数据，并计算每个类别的数量

```
# 建立一个字典 dict，用来计算每个类别的标签数量
train_dict = {}
# 读取整个训练数据集
for data in train_data:
    # 将读取到的标签转成 NumPy 格式
    label = data['label'].numpy()
    # 计算每个类别的数量：使用 dict 字典来计算每一个类别的数量
    train_dict[label] = train_dict.setdefault(label, 0) + 1
print(train_dict)
```

结果如下：

```
{0: 4492, 1: 4473, 2: 4491, 3: 4497, 4: 4481, 5: 4519, 6: 4509, 7: 4515,
 8: 4517, 9: 4506}
```

- 显示数据集部分的图像数据

```
# 建立一个显示图像的数组
output = np.zeros((32 * 8, 32 * 8, 3), dtype=np.uint8)
row = 0
# 每一次取 8 笔数据，共取 8 次，所以总共取得 64 笔数据
for data in train_data.batch(8).take(8):
    # 将获取的 8 笔数据堆叠起来，放入显示图像数组第 N 行中
    output[:, row*32:(row+1)*32] = np.vstack(data['image'].numpy())
    row += 1
# 设置显示窗口大小
plt.figure(figsize=(8, 8))
# 显示图像
plt.imshow(output)
```

结果如图 5-24 所示。

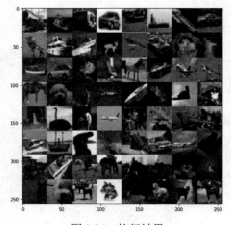

图 5-24　执行结果

Step 03 数据集设置。

- 数据预处理
  - 图像数据：将输入图像进行归一化，即将图像中的所有像素值除以255，让像素值缩放为0~1。
  - 标签数据：进行独热编码，例如类别2为[0, 0, 0, 0, 0, 0, 0, 0, 1, 0]。

```python
def parse_fn(dataset):
    # 图像归一化
    x = tf.cast(dataset['image'], tf.float32) / 255.
    # 将输出标签转换成独热编码
    y = tf.one_hot(dataset['label'], 10)
    return x, y
```

- 数据集设置

```python
AUTOTUNE = tf.data.experimental.AUTOTUNE  # 自动调整模式
batch_size = 64  # 批量大小
train_num = int(info.splits['train'].num_examples / 10) * 9  # 训练数据数量

# 打乱数据集
train_data = train_data.shuffle(train_num)
# 加载预处理 parse_fn function，CPU 数量为自动调整模式
train_data = train_data.map(map_func=parse_fn, num_parallel_calls=AUTOTUNE)
# 设置批量大小为 64，并启用 prefetch 模式（缓存空间为自动调整模式）
train_data = train_data.batch(batch_size).prefetch(buffer_size=AUTOTUNE)

# 加载预处理 parse_fn function，CPU 数量为自动调整模式
valid_data = valid_data.map(map_func=parse_fn, num_parallel_calls=AUTOTUNE)
# 设置批量大小为 64，并启用 prefetch 模式（缓存空间为自动调整模式）
valid_data = valid_data.batch(batch_size).prefetch(buffer_size=AUTOTUNE)

# 加载预处理 parse_fn function，CPU 数量为自动调整模式
test_data = test_data.map(map_func=parse_fn, num_parallel_calls=AUTOTUNE)
# 设置批量大小为 64，并启用 prefetch 模式（缓存空间为自动调整模式）
test_data = test_data.batch(batch_size).prefetch(buffer_size=AUTOTUNE)
```

Step 04 训练 Model-1（全连接神经网络）。

- 建立网络模型

这里使用了以下几种网络层：

  - keras.Input：输入层（输入图像的大小为 32×32×3）。
  - layers.Flatten：压平层（特征图转成一维张量）。
  - layers.Dropout：随机失活层（每次训练随机失活 30% 的神经元）。
  - layers.Dense：全连接层（隐藏层使用 ReLU 激活函数，输出层使用 Softmax 激活函数）。

```python
inputs = keras.Input(shape=(32, 32, 3))
x = layers.Flatten()(inputs)
```

```
x = layers.Dense(128, activation='relu')(x)
x = layers.Dropout(0.3)(x)
x = layers.Dense(256, activation='relu')(x)
x = layers.Dropout(0.3)(x)
x = layers.Dense(512, activation='relu')(x)
x = layers.Dropout(0.3)(x)
x = layers.Dense(512, activation='relu')(x)
x = layers.Dropout(0.3)(x)
x = layers.Dense(256, activation='relu')(x)
x = layers.Dropout(0.3)(x)
x = layers.Dense(64, activation='relu')(x)
x = layers.Dropout(0.3)(x)
outputs = layers.Dense(10, activation='softmax')(x)

# 建立网络模型（将输入到输出所有经过的网络层连接起来）
model_1 = keras.Model(inputs, outputs, name='model-1')
model_1.summary()    # 显示网络架构
```

结果如图 5-25 所示。

```
Model: "model-1"
_____
Layer (type)                 Output Shape              Param #
=================================================================
input_1 (InputLayer)         [(None, 32, 32, 3)]       0
_____
flatten (Flatten)            (None, 3072)              0
_____
dense (Dense)                (None, 128)               393344
_____
dropout (Dropout)            (None, 128)               0
_____
dense_1 (Dense)              (None, 256)               33024
_____
dropout_1 (Dropout)          (None, 256)               0
_____
dense_2 (Dense)              (None, 512)               131584
_____
dropout_2 (Dropout)          (None, 512)               0
_____
dense_3 (Dense)              (None, 512)               262656
_____
dropout_3 (Dropout)          (None, 512)               0
_____
dense_4 (Dense)              (None, 256)               131328
_____
dropout_4 (Dropout)          (None, 256)               0
_____
dense_5 (Dense)              (None, 64)                16448
_____
dropout_5 (Dropout)          (None, 64)                0
_____
dense_6 (Dense)              (None, 10)                650
=================================================================
Total params: 969,034
Trainable params: 969,034
Non-trainable params: 0
```

图 5-25　执行结果

- 创建模型存储的目录

```
model_dir = 'lab4-logs/models/'
os.makedirs(model_dir)
```

- 建立回调函数

```
# 将训练记录存成 TensorBoard 的记录文件
```

```
log_dir = os.path.join('lab4-logs', 'model-1')
model_cbk = keras.callbacks.TensorBoard(log_dir=log_dir)
# 存储最好的网络模型权重
model_mckp = keras.callbacks.ModelCheckpoint(model_dir + '/Best-model-1.h5',
                                  monitor='val_categorical_accuracy',
                                  save_best_only=True,
                                  mode='max')
```

- 设置训练使用的优化器、损失函数和评价指标函数

```
model_1.compile(keras.optimizers.Adam(),
         loss=keras.losses.CategoricalCrossentropy(),
         metrics=[keras.metrics.CategoricalAccuracy()])
```

- 训练网络模型

```
history_1 = model_1.fit(train_data,
              epochs=100,
              validation_data=valid_data,
              callbacks=[model_cbk, model_mckp])
```

结果如图 5-26 所示。

```
Epoch 95/100
704/704 [==============================] - 9s 13ms/step - loss: 1.1343 - categorical_accuracy: 0.5854 - val_loss: 2.1138 - val_categorical_accuracy: 0.4048
Epoch 96/100
704/704 [==============================] - 9s 13ms/step - loss: 1.1297 - categorical_accuracy: 0.5866 - val_loss: 2.0520 - val_categorical_accuracy: 0.4068
Epoch 97/100
704/704 [==============================] - 9s 12ms/step - loss: 1.1219 - categorical_accuracy: 0.5902 - val_loss: 2.1084 - val_categorical_accuracy: 0.4060
Epoch 98/100
704/704 [==============================] - 9s 12ms/step - loss: 1.1131 - categorical_accuracy: 0.5917 - val_loss: 2.1360 - val_categorical_accuracy: 0.4086
Epoch 99/100
704/704 [==============================] - 9s 13ms/step - loss: 1.1165 - categorical_accuracy: 0.5928 - val_loss: 2.0842 - val_categorical_accuracy: 0.3980
Epoch 100/100
704/704 [==============================] - 9s 13ms/step - loss: 1.1286 - categorical_accuracy: 0.5874 - val_loss: 2.0845 - val_categorical_accuracy: 0.3904
```

图 5-26　执行结果

Step 05 训练 Model-2（卷积神经网络）。

- 建立网络模型

这里使用了以下几种网络层：

> - keras.Input：输入层（输入图像的大小为 32×32×3）。
> - layers.Conv2D：卷积层（使用 ReLU 激活函数，以及 3×3 大小的 Kernel），这里 Kernel 是指卷积核。
> - layers.MaxPool2D：池化层（对特征图下采样）。
> - layers.Flatten：压平层（特征图转成一维张量）。
> - layers.Dropout：随机失活层（每次训练随机失活 50%的神经元）。
> - layers.Dense：全连接层（隐藏层使用 ReLU 激活函数，输出层使用 Softmax 激活函数）。

```
inputs = keras.Input(shape=(32, 32, 3))
```

```
x = layers.Conv2D(64, (3, 3), activation='relu')(inputs)
x = layers.MaxPool2D()(x)
x = layers.Conv2D(128, (3, 3), activation='relu')(x)
x = layers.Conv2D(256, (3, 3), activation='relu')(x)
x = layers.Conv2D(128, (3, 3), activation='relu')(x)
x = layers.Conv2D(64, (3, 3), activation='relu')(x)
x = layers.Flatten()(x)
x = layers.Dense(64, activation='relu')(x)
x = layers.Dropout(0.5)(x)
outputs = layers.Dense(10, activation='softmax')(x)

# 建立网络模型（将输入到输出所有经过的网络层连接起来）
model_2 = keras.Model(inputs, outputs, name='model-2')
model_2.summary()    # 显示网络架构
```

结果如图 5-27 所示。

```
Model: "model-2"
_____
Layer (type)                 Output Shape              Param #
=================================================================
input_2 (InputLayer)         [(None, 32, 32, 3)]       0
_____
conv2d (Conv2D)              (None, 30, 30, 64)        1792
_____
max_pooling2d (MaxPooling2D) (None, 15, 15, 64)        0
_____
conv2d_1 (Conv2D)            (None, 13, 13, 128)       73856
_____
conv2d_2 (Conv2D)            (None, 11, 11, 256)       295168
_____
conv2d_3 (Conv2D)            (None, 9, 9, 128)         295040
_____
conv2d_4 (Conv2D)            (None, 7, 7, 64)          73792
_____
flatten_1 (Flatten)          (None, 3136)              0
_____
dense_7 (Dense)              (None, 64)                200768
_____
dropout (Dropout)            (None, 64)                0
_____
dense_8 (Dense)              (None, 10)                650
=================================================================
Total params: 941,066
Trainable params: 941,066
Non-trainable params: 0
```

图 5-27　执行结果

- 建立回调函数

```
# 将训练记录存成 TensorBoard 的记录文件
log_dir = os.path.join('lab4-logs', 'model-2')
model_cbk = keras.callbacks.TensorBoard(log_dir=log_dir)
# 存储最好的网络模型权重
model_mckp = keras.callbacks.ModelCheckpoint(model_dir + '/Best-model-2.h5',
                                             monitor='val_categorical_accuracy',
                                             save_best_only=True, mode='max')
```

- 设置训练使用的优化器、损失函数和评价指标函数

```
model_2.compile(keras.optimizers.Adam(),
                loss=keras.losses.CategoricalCrossentropy(),
                metrics=[keras.metrics.CategoricalAccuracy()])
```

- 训练网络模型

```
history_2 = model_2.fit(train_data,
                        epochs=100,
                        validation_data=valid_data,
                        callbacks=[model_cbk, model_mckp])
```

结果如图 5-28 所示。

```
Epoch 95/100
704/704 [==============================] - 14s 20ms/step - loss: 0.1105 - categorical_accuracy: 0.9664 - val_loss: 2.7155
 - val_categorical_accuracy: 0.7102
Epoch 96/100
704/704 [==============================] - 14s 20ms/step - loss: 0.1375 - categorical_accuracy: 0.9604 - val_loss: 2.5822
 - val_categorical_accuracy: 0.6940
Epoch 97/100
704/704 [==============================] - 15s 21ms/step - loss: 0.1115 - categorical_accuracy: 0.9663 - val_loss: 2.8166
 - val_categorical_accuracy: 0.6986
Epoch 98/100
704/704 [==============================] - 18s 26ms/step - loss: 0.1080 - categorical_accuracy: 0.9665 - val_loss: 2.8064
 - val_categorical_accuracy: 0.6910
Epoch 99/100
704/704 [==============================] - 15s 21ms/step - loss: 0.1243 - categorical_accuracy: 0.9630 - val_loss: 2.7873
 - val_categorical_accuracy: 0.6956
Epoch 100/100
704/704 [==============================] - 15s 21ms/step - loss: 0.1188 - categorical_accuracy: 0.9635 - val_loss: 2.9477
 - val_categorical_accuracy: 0.6898
```

图 5-28 执行结果

Step 06 图像增强（Image Augmentation）。

- 读取图像：读取刚才显示的 8×8 图像数组中的一张图像，用于测试。

```
x = 3
y = 7
# 读取测试图像
image_test = output[y*32:(y+1)*32, x*32:(x+1)*32, :]
# 显示图像
plt.imshow(image_test)
```

结果如图 5-29 所示。

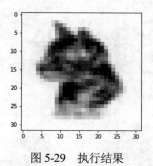

图 5-29 执行结果

- 水平翻转

```
def flip(x):
    """
    flip image(翻转图像)
    """
```

```
    x = tf.image.random_flip_left_right(x)   # 随机左右翻转图像
    return x

image_2 = flip(image_test)
# 将处理前后的图像进行水平合并
image = np.hstack((image_test, image_2))
# 显示图像
plt.imshow(image)
```

结果如图 5-30 所示。

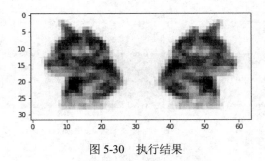

图 5-30　执行结果

- 颜色转换

```
def color(x):
    """
    Color change(改变颜色)
    """
    x = tf.image.random_hue(x, 0.08)              # 随机调整图像色调
    x = tf.image.random_saturation(x, 0.6, 1.6)   # 随机调整图像饱和度
    x = tf.image.random_brightness(x, 0.05)       # 随机调整图像亮度
    x = tf.image.random_contrast(x, 0.7, 1.3)     # 随机调整图像对比度
    return x

image_2 = color(image_test)
# 将处理前后的图像进行水平合并
image = np.hstack((image_test, image_2))
# 显示图像
plt.imshow(image)
```

结果如图 5-31 所示。

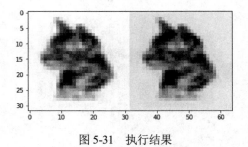

图 5-31　执行结果

- 图像旋转

```
def rotate(x):
    """
    Rotation image(图像旋转)
    """
    # 随机旋转 n 次（通过 minval 和 maxval 设置 n 的范围），每次旋转 90 度
    x = tf.image.rot90(x,tf.random.uniform(shape=[],minval=1,maxval=4,
                                    dtype=tf.int32))
    return x

image_2 = rotate(image_test)
# 将处理前后的图像进行水平合并
image = np.hstack((image_test, image_2))
# 显示图像
plt.imshow(image)
```

结果如图 5-32 所示。

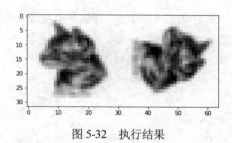

图 5-32　执行结果

- 图像缩放

```
def zoom(x, scale_min=0.6, scale_max=1.4):
    """
    Zoom Image(图像缩放)
    """
    h, w, c = x.shape
    scale = tf.random.uniform([], scale_min, scale_max)   # 随机缩放比例
    sh = h * scale   # 缩放后的图像长度
    sw = w * scale   # 缩放后的图像宽度
    x = tf.image.resize(x, (sh, sw))                      # 图像缩放
    x = tf.image.resize_with_crop_or_pad(x, h, w)         # 图像裁剪和填充
    return x

image_2 = zoom(image_test)
# 因为处理后图像会变成浮点数类型，所以需转换回来，否则显示时会出问题
image_2 = tf.cast(image_2, dtype=tf.uint8)
# 将处理前后的图像进行水平合并
image = np.hstack((image_test, image_2))
# 显示图像
plt.imshow(image)
```

结果如图 5-33 所示。

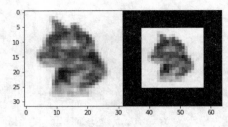

图 5-33 执行结果

Step 07 数据集设置（图像增强）。

- 重新加载数据集

前面已经设置过数据集，加载新的数据集就需要重新设置。

```
train_data = tfds.load("cifar10", split=train_split)
```

- 数据预处理
  - 归一化：将图像中的所有像素除以 255，即将像素值缩放到 0~1。
  - 图像增强：将图像进行水平翻转、旋转、缩放和颜色转换。
  - 标签数据：转换成独热编码，例如类别 2 为 [0, 0, 0, 0, 0, 0, 0, 0, 1, 0]。

```
def parse_aug_fn(dataset):
    """
    Image Augmentation(图像增强) function
    """
    x = tf.cast(dataset['image'], tf.float32) / 255.  # 图像归一化
    x = flip(x)  # 随机水平翻转
    # 触发颜色转换概率 50%
    x = tf.cond(tf.random.uniform([], 0, 1) > 0.5,lambda:color(x),lambda: x)
    # 触发图像旋转概率 0.25%
    x = tf.cond(tf.random.uniform([], 0, 1) > 0.75,lambda:rotate(x),lambda: x)
    # 触发图像缩放概率 50%
    x = tf.cond(tf.random.uniform([], 0, 1) > 0.5, lambda: zoom(x), lambda: x)
    # 将输出标签转成独热编码
    y = tf.one_hot(dataset['label'], 10)
    return x, y
```

- 数据集设置

```
# 打乱数据集
train_data = train_data.shuffle(train_num)
# 加载预处理 parse_aug_fn function，CPU 数量为自动调整模式
train_data = train_data.map(map_func=parse_aug_fn, num_parallel_calls=AUTOTUNE)
# 设置批量大小并启用 prefetch 预取模式（缓存空间为自动调整模式）
train_data = train_data.batch(batch_size).prefetch(buffer_size=AUTOTUNE)
```

- 测试数据集经过增强后的结果

```
# 由于前面已经将 train_data 的 batch size 设置为 64，因此取一次数据就有 64 笔
```

```
for images, labels in train_data.take(1):
    images = images.numpy()
# 建立一个显示图像的数组
output = np.zeros((32 * 8, 32 * 8, 3))
# 将 64 笔数据分别放入显示图像的数组
for i in range(8):
    for j in range(8):
        output[i*32:(i+1)*32, j*32:(j+1)*32, :] = images[i*8+j]
plt.figure(figsize=(8, 8))
# 显示图像
plt.imshow(output)
```

结果如图 5-34 所示。

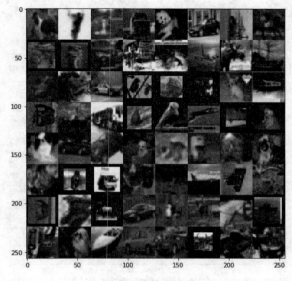

图 5-34　执行结果

Step 08　训练 Model-3（使用图像增强方法训练卷积神经网络）。

- 建立网络模型

这里使用了以下几种网络层：

- ➢ keras.Input：输入层（输入图像大小为 32×32×3）。
- ➢ layers.Conv2D：卷积层（使用 ReLU 激活函数，以及 3×3 大小的 Kernel），这里 Kernel 是指卷积核。
- ➢ layers.MaxPool2D：池化层（对特征图下采样）。
- ➢ layers.Flatten：压平层（特征图转成一维张量）。
- ➢ layers.Dropout：随机失活层（每次训练随机失活 50%的神经元）。
- ➢ layers.Dense：全连接层（隐藏层使用 ReLU 激活函数，输出层使用 Softmax 激活函数）。

```
inputs = keras.Input(shape=(32, 32, 3))
x = layers.Conv2D(64, (3, 3), activation='relu')(inputs)
```

```
x = layers.MaxPool2D()(x)
x = layers.Conv2D(128, (3, 3), activation='relu')(x)
x = layers.Conv2D(256, (3, 3), activation='relu')(x)
x = layers.Conv2D(128, (3, 3), activation='relu')(x)
x = layers.Conv2D(64, (3, 3), activation='relu')(x)
x = layers.Flatten()(x)
x = layers.Dense(64, activation='relu')(x)
x = layers.Dropout(0.5)(x)
outputs = layers.Dense(10, activation='softmax')(x)

# 建立网络模型（将输入到输出所有经过的网络层连接起来）
model_3 = keras.Model(inputs, outputs, name='model-3')
# 显示网络架构
model_3.summary()
```

结果如图 5-35 所示。

```
Model: "model-3"
_____
Layer (type)                 Output Shape              Param #
=================================================================
input_3 (InputLayer)         [(None, 32, 32, 3)]       0
conv2d_5 (Conv2D)            (None, 30, 30, 64)        1792
max_pooling2d_1 (MaxPooling2 (None, 15, 15, 64)        0
conv2d_6 (Conv2D)            (None, 13, 13, 128)       73856
conv2d_7 (Conv2D)            (None, 11, 11, 256)       295168
conv2d_8 (Conv2D)            (None, 9, 9, 128)         295040
conv2d_9 (Conv2D)            (None, 7, 7, 64)          73792
flatten_2 (Flatten)          (None, 3136)              0
dense_9 (Dense)              (None, 64)                200768
dropout_1 (Dropout)          (None, 64)                0
dense_10 (Dense)             (None, 10)                650
=================================================================
Total params: 941,066
Trainable params: 941,066
Non-trainable params: 0
```

图 5-35　执行结果

- 建立回调函数

```
# 存储训练记录文件
log_dir = os.path.join('lab4-logs', 'model-3')
model_cbk = keras.callbacks.TensorBoard(log_dir=log_dir)
# 将训练记录存成 TensorBoard 的记录文件
model_mckp = keras.callbacks.ModelCheckpoint(model_dir + '/Best-model-3.h5',
                                             monitor='val_categorical_accuracy',
                                             save_best_only=True,
                                             mode='max')
```

- 设置训练使用的优化器、损失函数和评价指标函数

```
model_3.compile(keras.optimizers.Adam(),
                loss=keras.losses.CategoricalCrossentropy(),
```

```
                    metrics=[keras.metrics.CategoricalAccuracy()])
```

- 训练网络模型

```
history_3 = model_3.fit(train_data,
                epochs=100,
                validation_data=valid_data,
                callbacks=[model_cbk, model_mckp])
```

结果如图 5-36 所示。

```
704/704 [==============================] - 13s 19ms/step - loss: 0.8218 - categorical_accuracy: 0.7270 - val_loss: 0.6485
 - val_categorical_accuracy: 0.7930
Epoch 96/100
704/704 [==============================] - 13s 19ms/step - loss: 0.8061 - categorical_accuracy: 0.7303 - val_loss: 0.6787
 - val_categorical_accuracy: 0.7776
Epoch 97/100
704/704 [==============================] - 14s 19ms/step - loss: 0.8112 - categorical_accuracy: 0.7307 - val_loss: 0.6559
 - val_categorical_accuracy: 0.7860
Epoch 98/100
704/704 [==============================] - 13s 19ms/step - loss: 0.8126 - categorical_accuracy: 0.7310 - val_loss: 0.6569
 - val_categorical_accuracy: 0.7852
Epoch 99/100
704/704 [==============================] - 15s 21ms/step - loss: 0.8048 - categorical_accuracy: 0.7290 - val_loss: 0.6193
 - val_categorical_accuracy: 0.8024
Epoch 100/100
704/704 [==============================] - 15s 22ms/step - loss: 0.8059 - categorical_accuracy: 0.7328 - val_loss: 0.6620
 - val_categorical_accuracy: 0.7866
```

图 5-36 执行结果

Step 09 比较 3 种网络的训练结果。

- 读取各自最佳的网络权重

```
model_1.load_weights('lab4-logs/models/Best-model-1.h5')
model_2.load_weights('lab4-logs/models/Best-model-2.h5')
model_3.load_weights('lab4-logs/models/Best-model-3.h5')
```

- 在测试集数据上验证

```
loss_1, acc_1 = model_1.evaluate(test_data)
loss_2, acc_2 = model_2.evaluate(test_data)
loss_3, acc_3 = model_3.evaluate(test_data)
```

- 显示预测的损失值与正确率结果

从训练结果可以观察到加入图像增强方法的卷积神经网络效果最好，其次是卷积神经网络，最后才是全连接神经网络。

```
loss = [loss_1, loss_2, loss_3]
acc = [acc_1, acc_2, acc_3]
dict = {"Accuracy": acc, "Loss": loss}
pd.DataFrame(dict)
```

结果如图 5-37 所示。

	Loss	Accuracy
0	1.662437	0.4453
1	2.209503	0.7211
2	0.631140	0.7980

图 5-37　执行结果

- 启动 TensorBoard（命令行）查看训练记录

```
tensorboard --logdir lab4-logs
```

首先，观察训练数据的历史曲线，如图 5-38 所示，各个曲线分别代表：

> Model-1：全连接神经网络。
> Model-2：卷积神经网络。
> Model-3：卷积神经网络加入图像增强方法进行训练。

由此曲线图可以发现，在训练数据上 Model-2 > Model-3 > Model-1。

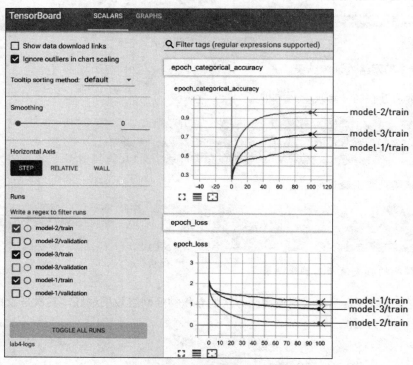

图 5-38　训练数据的历史曲线图

再来观察验证数据的历史曲线，如图 5-39 所示。

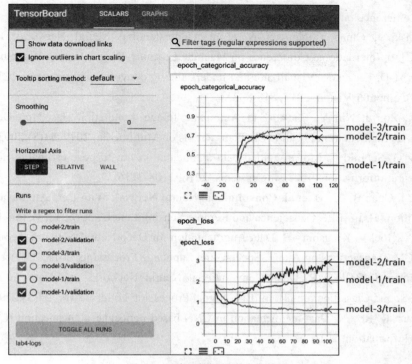

图 5-39　验证数据的历史曲线图

- Model-1：全连接神经网络。
- Model-2：卷积神经网络。
- Model-3：卷积神经网络加入图像增强方法进行训练。

由此曲线图可以发现，在验证数据上，Model-3→Model-2→Model-1，并且从曲线变化发现 Model-2 有很严重的过拟合问题，而使用图像增强的网络模型 Model-3 可以改善过拟合问题。

5.4　参考文献

[1] Krizhevsky A, Sutskever I, Hinton G E. Imagenet classification with deep convolutional neural networks [C]. Advances in Neural Information Processing Systems, 2012, 1097-1105.

[2] Deng J, Dong W, Socher R, et al. Imagenet: A large-scale hierarchical image database [C]. Proceedings of the IEEE Conference on Computer Vision and Pattern Recognition, 2009, 248-255.

[3] Wikipedia. Kernel. https://en.wikipedia.org/wiki/Kernel_(image_processing).

[4] Mahendran A, Vedaldi A. Understanding image representations by measuring their equivariance and equivalence [C]. Proceedings of the IEEE Conference on Computer Vision and Pattern Recognition, 2015, 5188-5196.

[5] Zhou B, Khosla A, Lapedriza A, et al. Object detectors emerge in deep scene cnns [C]. In Proc.

International Conference on Learning Representations, 2014.

[6] Yosinski J, Clune J, Nguyen A, et al. Understanding Neural Networks Through Deep Visualization[C]. In International Conference on Machine Learning-Deep Learning Workshop, 2015, 12.

[7] Zeiler M D, Fergus R. Visualizing and understanding convolutional networks [C]. In European Conference on Computer Vision, 2014, 818-833.

[8] Ke X, Zou J, Niu Y. End-to-End Automatic Image Annotation Based on Deep CNN and Multi-Label Data Augmentation [J]. In IEEE Transactions on Multimedia, 2019, 21(8):2093-2106.

[9] Rogez G, Schmid C. Mocap-guided data augmentation for 3d pose estimation in the wild [J]. Advances in neural information processing systems, 2016, 3108-3116.

[10] Ding J, Chen B, Liu H, et al. Convolutional Neural Network With Data Augmentation for SAR Target Recognition [J]. In IEEE Geoscience and Remote Sensing Letters, 2016, 13(3):364-368.

[11] Cui X, Goel V, Kingsbury B. Data Augmentation for Deep Neural Network Acoustic Modeling [J]. in IEEE/ACM Transactions on Audio, Speech, and Language Processing, 2015, 23(9):1469-1477.

[12] Salamon J, Bello J P. Deep Convolutional Neural Networks and Data Augmentation for Environmental Sound Classification [J]. in IEEE Signal Processing Letters, 2017, 24(3):279-283.

[13] Dellana R, Roy K. Data augmentation in CNN-based periocular authentication [C]. International Conference on Information, 2016.

第 6 章

神经网络训练技巧

学习目标

- 掌握神经网络反向传播原理
- 了解权重默认值的重要性
- 学会使用 TensorBoard 观察每一层网络权重的分布
- 了解批量归一化的重要性
- 总结各种训练方法所带来的效益

6.1 反向传播

在介绍神经网络训练技巧之前,需要先理解神经网络如何更新权重。本节介绍的反向传播(Back Propagation,BP)[1]是一种结合梯度下降法更新神经网络权重的方法。首先,回顾一下第 3 章介绍的梯度下降法,梯度下降法为常用的优化算法,通过计算梯度并沿着梯度反方向移动,进而减少网络预测值与样本标注值之间的误差。权重更新公式如下:

$$W = W - \eta \frac{\partial L}{\partial W}$$

L:损失函数。
η:学习率。
W:权重。

通过以上式子,在每一次迭代中,使用反向传播计算损失函数对每个 W 权重的导数。下面以简单的网络模型例子来讲解 $\frac{\partial L}{\partial W}$ 是如何计算的。

反向传播范例

下面为单层神经网络正向传播（Forward Pass）的公式和示意图，如图 6-1 所示。

$$z = x_1w_1 + x_2w_2 + x_3w_3$$

$$\hat{y} = f(z)$$

$$L = (y - \hat{y})^2$$

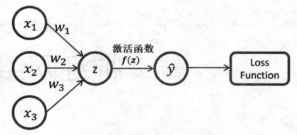

图 6-1　单层神经网络正向传播示意图

如图 6-2 所示，通过反向传播法计算 $\frac{\partial L}{\partial w_1}$、$\frac{\partial L}{\partial w_2}$ 和 $\frac{\partial L}{\partial w_3}$，计算过程如下：

$$\frac{\partial L}{\partial w_1} = \frac{\partial L}{\partial \hat{y}} \frac{\partial \hat{y}}{\partial z} \frac{\partial z}{\partial w_1} = -2(y - \hat{y}) \times f'(z) \times x_1$$

$$\frac{\partial L}{\partial w_2} = \frac{\partial L}{\partial \hat{y}} \frac{\partial \hat{y}}{\partial z} \frac{\partial z}{\partial w_2} = -2(y - \hat{y}) \times f'(z) \times x_2$$

$$\frac{\partial L}{\partial w_3} = \frac{\partial L}{\partial \hat{y}} \frac{\partial \hat{y}}{\partial z} \frac{\partial z}{\partial w_3} = -2(y - \hat{y}) \times f'(z) \times x_3$$

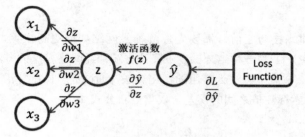

图 6-2　单层神经网络反向传播示意图

表 6-1 列出了 3 种常见的激活函数，其中 Sigmoid 和 Tanh 作为神经网络隐藏层的激活函数，都有可能造成梯度消失（Vanishing Gradient）问题。以 Sigmoid 为例来解释，反向传播每经过一层 Sigmoid 激活函数，都需乘上 Sigmoid 的导函数，而 Sigmoid 导函数最大值为 $f'(0) = 0.25$，代表经过一次就会衰减 0.25 倍，如果神经网络为 5 层，那么梯度至少会被缩减 0.25^5（0.25 的 5 次方），如此将造成神经网络前面几层的权重很难被更新，同理 Tanh 也有相同的情况。所以现在大多数隐藏层激活函数都使用 ReLU，另外 ReLU 的计算速度比 Sigmoid 和 Tanh 要快上许多。

表 6-1 激活函数公式整理

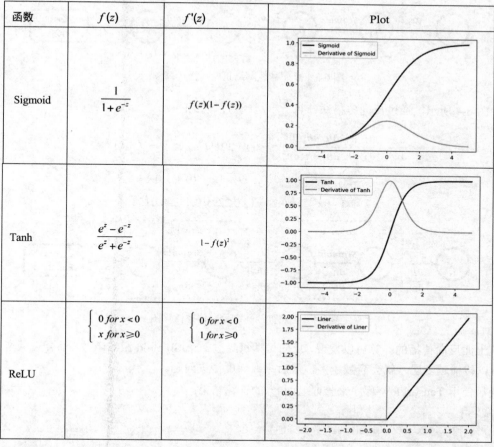

Sigmoid 激活函数的梯度消失范例

此范例使用两层的神经网络,每一层网络层都会使用 Sigmoid 激活函数(f),并且已经知道输入(x=0.4)、权重(w_1=1、w_2=1)和预期输出值(y=1),图 6-3 所示为网络层正向传播的公式和示意图。

$$h_1 = xw_1 \cong 0.4$$

$$h_2 = f(h_1) \cong 0.599$$

$$h_3 = h_2 w_2 \cong 0.599$$

$$\hat{y} = f(h_3) \cong 0.645$$

$$L = (y - \hat{y})^2 \cong 0.126$$

$$f(x) = \frac{1}{1+e^{-x}}$$

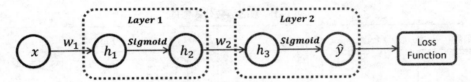

图 6-3 单层神经网络正向传播计算图

如图 6-4 所示，通过反向传播法计算 $\frac{\partial L}{\partial w_1}$，计算过程如下：

$$\frac{\partial L}{\partial w_1} = \frac{\partial L}{\partial \hat{y}} \frac{\partial \hat{y}}{\partial h_3} \frac{\partial h_3}{\partial h_2} \frac{\partial h_2}{\partial h_1} \frac{\partial h_1}{\partial w_1} = -2(y-\hat{y}) \times f'(h_3) \times w_2 \times f'(h_1) \times x =$$

$$-2(1-0.645) \times f'(0.599) \times 1 \times f'(0.4) \times 0.4 =$$

$$-2(1-0.645) \times 0.24 \times 1 \times 0.229 \times 0.4 = -0.0156$$

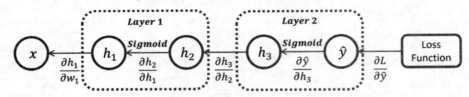

图 6-4 单层神经网络反向传播计算图

从上面反向传播的计算可以发现，反向传播每经过一个 Sigmoid 激活函数，都会衰减至少 0.25 倍，所以传播越多层，就会衰减越多，最后导致梯度消失问题。

最后使用 TensorFlow 程序来验证上面 $\frac{\partial L}{\partial w_1}$ 的计算结果：

```python
import tensorflow as tf
# 声明输入、权重和预测输出值
x = 0.4
w1, w2 = tf.Variable(1.0), tf.Variable(1.0)
y = 1

# 正向传播会被记录到 tape 中
with tf.GradientTape() as tape:
    # 正向传播的计算
    h1 = x * w1
    h2 = tf.sigmoid(h1)
    h3 = h2 * w2
    y_hat = tf.sigmoid(h3)
    loss = (y - y_hat)**2
# 反向传播 tape 计算权重 w1 梯度
gradients = tape.gradient(loss, w1)
print(gradients)
```

结果如下：

```
tf.Tensor(-0.015601176, shape=(), dtype=float32)
```

6.2 权重初始化

神经网络的权重初始化对网络训练非常重要，会直接影响训练是否成功[2]-[5]。接下来介绍 3 种不同的权重初始化方法，并分析权重初始化方法对网络模型的影响。下面为 3 种不同权重初始化的简介：

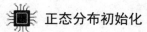 正态分布初始化

下面使用 Sigmoid 激活函数分析标准差为 1 的正态分布和标准差为 0.01 的正态分布初始化权重的结果，最终分析结果显示标准差为 1 的正态分布会造成梯度消失。

 Xavier/Glorot 初始化[6]

改善使用 Sigmoid 激活函数在标准差为 0.01 的正态分布初始化中遇到的问题，让输出值不会全部集中在 0.5，但在使用 ReLU 激活函数的情况下，可能会发生梯度消失。

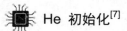 He 初始化[7]

改善使用 ReLU 激活函数在 Xavier/Glorot 初始化中遇到的问题，让每一层的输出值分布很平均。

导入下面的范例程序必要的套件：

```
import numpy as np
import tensorflow as tf
import matplotlib.pyplot as plt
from tensorflow import keras
from tensorflow.keras import layers
from tensorflow.keras import initializers
```

6.2.1 正态分布

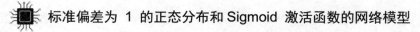 标准偏差为 1 的正态分布和 Sigmoid 激活函数的网络模型

- 建立网络模型

这里使用了以下几种网络层：

> keras.Input：输入层（输入大小为 100 维向量）。
> layers.Dense：全连接层（激活函数使用 Sigmoid，不使用 Bias，并使用标准差为 1 的正态分布初始化权重）。

```
inputs = keras.Input(shape=(100,))
x1 = layers.Dense(100, 'sigmoid', False,
```

```
                    initializers.RandomNormal(0, 1))(inputs)
x2 = layers.Dense(100, 'sigmoid', False, initializers.RandomNormal(0, 1))(x1)
x3 = layers.Dense(100, 'sigmoid', False, initializers.RandomNormal(0, 1))(x2)
x4 = layers.Dense(100, 'sigmoid', False, initializers.RandomNormal(0, 1))(x3)
x5 = layers.Dense(100, 'sigmoid', False, initializers.RandomNormal(0, 1))(x4)
model_1 = keras.Model(inputs, [x1, x2, x3, x4, x5])
```

- 显示每层网络输出值的分布

```
x = np.random.randn(100, 100)
outputs = model_1.predict(x)
for i, layer_output in enumerate(outputs):
    plt.subplot(1, 5, i+1)              # 选择显示在表中的哪个格子里
    plt.title(str(i+1) + "-layer")      # 设置直方图的标题
    if i != 0: plt.yticks([], [])       # 只显示第一列直方图的 y 轴
    plt.hist(layer_output.flatten(), 30, range=[0,1])  # 画出直方图
plt.show()
```

结果如图 6-5 所示。

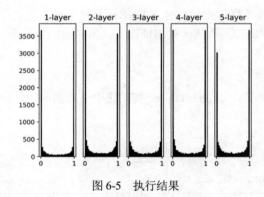

图 6-5　执行结果

从各层的输出值分布图可以观察到，大部分的输出值都落在 0 与 1 附近，但在计算反向传播时，传播值会趋近于 0（对照表 6-1 的 Sigmoid 函数），造成梯度消失的问题。因此，标准差为 1 的正态分布并不是很好的权重初始化方法。

标准偏差为 0.01 的正态分布和 Sigmoid 激活函数的网络模型

- 建立网络模型

这里使用了以下几种网络层：

> keras.Input：输入层（输入大小为 100 维向量）。
> layers.Dense：全连接层（激活函数使用 Sigmoid，不使用 Bias，并使用标准差为 0.01 的正态分布初始化权重）。

```
inputs = keras.Input(shape=(100,))
x1 = layers.Dense(100, 'sigmoid', False,
                  initializers.RandomNormal(0, 0.01))(inputs)
x2 = layers.Dense(100, 'sigmoid', False, initializers.RandomNormal(0, 0.01))(x1)
```

```
    x3 = layers.Dense(100, 'sigmoid', False, initializers.RandomNormal(0, 0.01))(x2)
    x4 = layers.Dense(100, 'sigmoid', False, initializers.RandomNormal(0,
0.01))(x3)
    x5 = layers.Dense(100, 'sigmoid', False, initializers.RandomNormal(0,
0.01))(x4)
    model_2 = keras.Model(inputs, [x1, x2, x3, x4, x5])
```

- 显示每层网络输出值的分布

```
x = np.random.randn(100, 100)
outputs = model_2.predict(x)
for i, layer_output in enumerate(outputs):
    plt.subplot(1, 5, i+1)              # 选择显示在表中的哪个格子里
    plt.title(str(i+1) + "-layer")      # 设置直方图的标题
    if i != 0: plt.yticks([], [])       # 只显示第一列直方图的 y 轴
    plt.hist(layer_output.flatten(), 30, range=[0,1])  # 画出直方图
plt.show()
```

结果如图 6-6 所示。

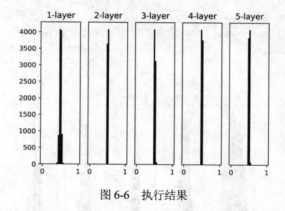

图 6-6　执行结果

从上面的执行结果可以发现，第二层之后的输出值大多分布在 0.5 附近，在计算反向传播时，传播值会落在 0.25 附近（对照表 6-1 的 Sigmoid 函数），虽然改善了标准差为 1 的正态分布初始化方法的梯度消失问题，但如果构建太多层网络，还是有可能发生梯度消失问题。

6.2.2　Xavier/Glorot 初始化

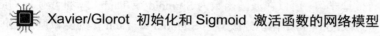 Xavier/Glorot 初始化和 Sigmoid 激活函数的网络模型

Xavier 初始化（又称作 Glorot 初始化）[6]，是由 Xavier Glorot 和 Bengio 于 2010 年提出的权重初始化方法，该方法被广泛用于许多深度学习框架，并被当作默认的权重初始化方法，Keras 也使用此方法作为网络权重的默认初始化方法。

- 建立网络模型

这里使用了以下几种网络层：

➢ keras.Input: 输入层（输入大小为 100 维向量）。
➢ layers.Dense: 全连接层（激活函数使用 Sigmoid, 不使用 Bias, 并使用 Glorot 初始化权重）。

```
inputs = keras.Input(shape=(100,))
x1 = layers.Dense(100, 'sigmoid', False,
initializers.glorot_normal())(inputs)
x2 = layers.Dense(100, 'sigmoid', False, initializers.glorot_normal())(x1)
x3 = layers.Dense(100, 'sigmoid', False, initializers.glorot_normal())(x2)
x4 = layers.Dense(100, 'sigmoid', False, initializers.glorot_normal())(x3)
x5 = layers.Dense(100, 'sigmoid', False, initializers.glorot_normal())(x4)
model_3 = keras.Model(inputs, [x1, x2, x3, x4, x5])
```

- 显示每层网络输出值的分布

```
x = np.random.randn(100, 100)
outputs = model_3.predict(x)
for i, layer_output in enumerate(outputs):
    plt.subplot(1, 5, i+1)            # 选择显示在表中的哪个格子里
    plt.title(str(i+1) + "-layer")    # 设置直方图的标题
    if i != 0: plt.yticks([], [])     # 只显示第一列直方图的 y 轴
    plt.hist(layer_output.flatten(), 30, range=[0,1])  # 画出直方图
plt.show()
```

结果如图 6-7 所示。

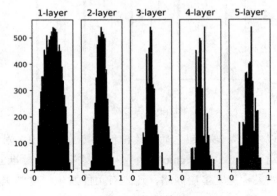

图 6-7　执行结果

从结果来看，相对于标准差为 0.01 的正态分布初始化，Glorot 初始化的输出值分布较广，提升了网络的多样性，进而可以有效地学习，所以在激活函数为 Sigmoid 或 Tanh 时，推荐使用 Glorot 初始化来对权重进行初始化。虽然 Glorot 初始化已经改善了 Sigmoid 或 Tanh 作为隐藏层面临的困境，但是仍存在反向传播会衰减的问题，所以隐藏层的激活函数还是推荐使用 ReLU。接下来介绍适合 ReLU 的权重初始化方法。

 Xavier/Glorot 初始化和 ReLU 激活函数的网络模型

- 建立网络模型

这里使用了以下几种网络层：

> keras.Input：输入层（输入大小为 100 维向量）。

> layers.Dense：全连接层（激活函数使用 ReLU，不使用 Bias，并使用 Glorot 初始化权重）。

```
inputs = keras.Input(shape=(100,))
x1 = layers.Dense(100, 'relu', False, initializers.glorot_normal())(inputs)
x2 = layers.Dense(100, 'relu', False, initializers.glorot_normal())(x1)
x3 = layers.Dense(100, 'relu', False, initializers.glorot_normal())(x2)
x4 = layers.Dense(100, 'relu', False, initializers.glorot_normal())(x3)
x5 = layers.Dense(100, 'relu', False, initializers.glorot_normal())(x4)
model_4 = keras.Model(inputs, [x1, x2, x3, x4, x5])
```

- 显示每层网络输出值的分布

```
x = np.random.randn(100, 100)
outputs = model_4.predict(x)
for i, layer_output in enumerate(outputs):
    plt.subplot(1, 5, i+1)         # 选择显示在表中的哪个格子里
    plt.title(str(i+1) + "-layer")  # 设置直方图的标题
    if i != 0: plt.yticks([], [])   # 只显示第一列直方图的 y 轴
    plt.hist(layer_output.flatten(), 30, range=[0,1])  # 画出直方图
plt.show()
```

结果如图 6-8 所示。

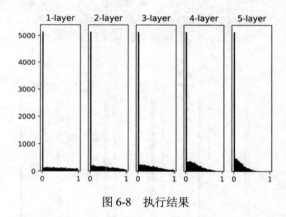

图 6-8　执行结果

从结果图可以观察到，随着层数的增加，输出值的分布会越来越靠近 0，如此很有可能造成梯度消失的问题，因此使用 ReLU 激活函数和 Glorot 初始化时并不能搭建太深的网络模型。

6.2.3　He 初始化

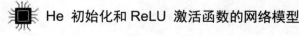　He 初始化和 ReLU 激活函数的网络模型

He 初始化[7]由 Kaiming He 在 2015 年提出，是 Glorot 初始化的变形，可以解决 Glorot 初始化用于 ReLU 激活函数时梯度消失的问题。

- 建立网络模型

这里使用了以下几种网络层：

> keras.Input：输入层（输入大小为 100 维向量）。
> layers.Dense：全连接层（激活函数使用 ReLU，不使用 Bias，并使用 He 初始化权重）。

```
inputs = keras.Input(shape=(100,))
x1 = layers.Dense(100, 'relu', False, initializers.he_normal())(inputs)
x2 = layers.Dense(100, 'relu', False, initializers.he_normal())(x1)
x3 = layers.Dense(100, 'relu', False, initializers.he_normal())(x2)
x4 = layers.Dense(100, 'relu', False, initializers.he_normal())(x3)
x5 = layers.Dense(100, 'relu', False, initializers.he_normal())(x4)
model_5 = keras.Model(inputs, [x1, x2, x3, x4, x5])
```

- 显示每层网络输出值的分布

```
x = np.random.randn(100, 100)
outputs = model_5.predict(x)
for i, layer_output in enumerate(outputs):
    plt.subplot(1, 5, i+1)             # 选择显示在表中的哪个格子里
    plt.title(str(i+1) + "-layer")     # 设置直方图的标题
    if i != 0: plt.yticks([], [])      # 只显示第一列直方图的 y 轴
    plt.hist(layer_output.flatten(), 30, range=[0,1])  # 画出直方图
plt.show()
```

结果如图 6-9 所示。

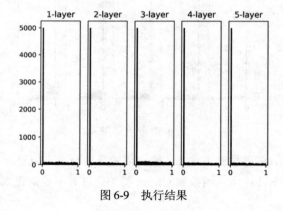

图 6-9　执行结果

从输出结果图来看，使用 He 初始化方法成功解决了使用 Glorot 初始化方法所造成的输出分布会聚集到 0 附近的问题，尤其网络越深，Glorot 初始化方法越会聚集到 0，由于 He 初始化方法每一层的输出分布都很平均，因此成功解决了 Glorot 初始化的问题。

最后总结这一节学习到的内容：

- 使用 Sigmoid 或 Tanh 作为隐藏层激活函数，建议使用 Glorot 初始化（又叫作 Xavier 初始化）。
- 使用 ReLU 或其他变形作为隐藏层的激活函数，建议使用 He 初始化。
- 使用 ReLU 作为隐藏层激活函数，其性能优于 Sigmoid 或 Tanh。

6.3 批量归一化

在 6.2 节说明了权重默认值的重要性,使用不同的权重初始化技巧让网络各层的输出值得以平均分布,进而让神经网络学习到更多样的特征。这一节所说明的批量归一化(Batch Normalization)就是强制让每一层的输出值分布均匀。

6.3.1 批量归一化介绍

批量归一化[8]是 Google 公司于 2015 年提出的对网络层输出值进行归一化的算法,它不仅可以让网络的收敛速度加快,还可以在一定程度缓解梯度消失和梯度爆炸的问题,进而让网络训练更加容易和稳定。现在大部分网络架构都会使用批量归一化,基本上已经成为必要的方法,后续还有各种归一化的方法陆续发表,例如:LayerNormalization[9]、InstanceNorm[10]、GroupNorm[11]和 SwitchableNorm[12]。

批量归一化的核心思想是对输入网络的每一批数据、每一层网络输出值都进行归一化。至于为什么要对每一层输出值都进行归一化,例如图 6-10(a)网络层输出值平均分布在-10 和 10 之间,经过 Tanh 激活函数后输出值会集中在-1 和 1,因此造成梯度消失的问题(对照表 6-1 的 Tanh 函数);而图 6-10(b)将网络层的输出值进行归一化后,输出值分布在-2 和 2 之间,再经过 Tanh 激活函数后的输出值分布会更为平均,并且不会全部集中在-1 和 1,进而改善了梯度消失的问题。

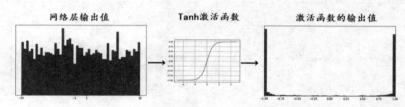

(a)网络层输出值直接进入 Tanh 激活函数

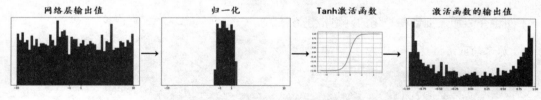

(b)网络层输出值进行归一化,再进入 Tanh 激活函数

图 6-10 网络层输出值进行归一化与没进行归一化的比较图

批量归一化除了解决梯度消失问题之外,还解决了内部协变量偏移(Internal Covariate Shift)问题。内部协变量偏移问题是指:当神经网络更新权重时,网络层的输出值分布会产生变化,因为输出值分布的变化,后一层网络要不断地适应前一层的输出值分布变化,即每一层都需要根据前一层的输出值分布进行调整,由于各层之间是环环相扣的,因此其中一层变化太大,后面跟不上,训

练就会出现问题。为了解决这个问题，之前的做法是将学习率（Learning Rate）设置得小一点，例如 10^{-3}，而批量归一化方法是让输出值分布较固定，因此可以缓解内部协变量偏移的问题，此外也可以使用较大的学习率来训练网络，以加速网络训练速度。

以下整理批量归一化的几个优点：

- 不需要依赖权重初始化方法。
- 减少梯度消失（Vanishing Gradient）或梯度爆炸（Exploding Gradient）。
- 避免过拟合（可以减少 Dropout 或 Regularization 的使用）。
- 加快学习速度（减少内部协变量偏移的问题，因而可以使用较大的学习率进行训练）。

> **说　明**
>
> 批量归一化的原理与归一化方法很接近，都要计算平均值和标准偏差，但是网络训练时每一批的平均值和标准偏差都会变动，因此每一批训练都会计算一次平均值和标准偏差并累加起来。另外，还有对输出进行缩放（Scale）和偏移（Shift）的参数，更多的原理细节可以参考 Batch Normalization: Accelerating Deep Network Training by Reducing Internal Covariate Shift[8]论文。

6.3.2　批量归一化网络架构

批量归一化的网络架构与前面几个章节介绍的网络有些不一样，主要是批量归一化网络层的位置大部分都介于卷积层与激活函数之间，如图 6-11 所示。

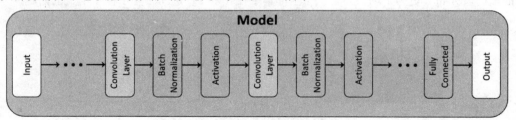

图 6-11　卷积神经网络加入批量归一化网络架构

下面将介绍加入批量归一化的网络搭建方法。

- 原来的网络搭建

```
……省略
x = layers.Conv2D(128, (3, 3), activation='relu')(x)
……省略
```

- 加入批量归一化的网络搭建

```
……省略
x = layers.Conv2D(128, (3, 3))(x)
x = layers.BatchNormalization()(x)
x = layers.ReLU()(x)
……省略
```

6.4 实验一：使用 CIFAR-10 数据集实验 3 种权重初始化方法

本章会继续 5.3 节的范例程序，在 CIFAR-10 数据集上训练 3 种不同权重初始化方法的网络模型（每个模型的网络大小都与 5.3.2 小节中的卷积神经网络相同，并使用 ReLU 激活函数），这 3 种初始化方法分别为正态分布（std0.01）初始化、Glorot 初始化和 He 初始化，最后比较这 3 种权重初始化方法训练的正确率。

6.4.1 新建项目

建议使用 Jupyter Notebook 来执行本小节的程序代码，操作流程如下：

Step 01 启动 Jupyter Notebook。

在 Terminal（Ubuntu）或命令提示符（Windows）中输入如下指令：

```
jupyter notebook
```

Step 02 新建执行文件。

单击界面右上角的 New 下拉按钮，然后单击所安装的 Python 解释器（在 Jupyter 中都称为 Kernel）来启动它，如图 6-12 所示，显示了 3 个不同的 Kernel：

- Python 3：本地端 Python。
- tf2：虚拟机 Python（TensorFlow-cpu 版本）。
- tf2-gpu：虚拟机 Python（TensorFlow-gpu 版本）。

图 6-12　新建执行文件

Step 03 执行程序代码。

按 Shift + Enter 快捷键执行单行程序代码，如图 6-13 所示。

图 6-13　Jupyter 环境界面

接下来,后续的程序代码都可在 Jupyter Notebook 上执行。

6.4.2　建立图像增强函数

由于之后很多章节都会用到图像增强函数,因此本小节建立一个 Python 文件,里面包含图像增强函数,之后再次使用图像增强函数时直接导入这个文件即可。

Step 01 新建文件。

在 Jupyter 中新建文件,如图 6-14 所示。

图 6-14　新建文件

Step 02 重新命名。

将文件名修改成 preprocessing.py,如图 6-15 所示。

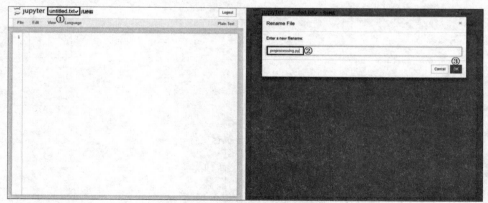

图 6-15　修改文件名

Step 03 导入必要的套件。

```
import tensorflow as tf
```

Step 04 建立图像增强函数。

在 preprocessing.py 文件中加入图像增强函数。

```
def flip(x):
    """
    flip image(翻转图像)
    """
    x = tf.image.random_flip_left_right(x)      # 随机左右翻转图像
    return x

def color(x):
    """
    Color change(改变颜色)
    """
    x = tf.image.random_hue(x, 0.08)             # 随机调整图像色调
    x = tf.image.random_saturation(x, 0.6, 1.6)  # 随机调整图像饱和度
    x = tf.image.random_brightness(x, 0.05)      # 随机调整图像亮度
    x = tf.image.random_contrast(x, 0.7, 1.3)    # 随机调整图像对比度
    return x

def rotate(x):
    """
    Rotation image(图像旋转)
    """
    # 随机旋转 n 次（通过 minval 和 maxval 设置 n 的范围），每次旋转 90 度
    x = tf.image.rot90(x,tf.random.uniform(shape=[],minval=1,maxval=4,
                       dtype=tf.int32))
    return x

def zoom(x, scale_min=0.6, scale_max=1.4):
    """
    Zoom Image(图像缩放)
    """
    h, w, c = x.shape
    scale = tf.random.uniform([], scale_min, scale_max)  # 随机缩放比例
    sh = h * scale                                        # 缩放后的图像长度
    sw = w * scale                                        # 缩放后的图像宽度
    x = tf.image.resize(x, (sh, sw))                      # 图像缩放
    x = tf.image.resize_with_crop_or_pad(x, h, w)         # 图像裁剪和填充
    return x
```

Step 05 数据预处理（Data Preprocessing）。

在 preprocessing.py 文件中加入数据预处理函数。

- 训练数据
  - 归一化:将图像中的所有像素除以 255,即将像素值缩放到 0~1。
  - 图像增强:将图像进行水平翻转、旋转、缩放和颜色转换。
  - 标签数据:转换成独热编码,例如类别 2 为[0, 0, 0, 0, 0, 0, 0, 0, 1, 0]。
- 验证和测试数据
  - 归一化:将图像中的所有像素除以 255,即将像素值缩放到 0~1。
  - 标签数据:转换成独热编码,例如类别 2 为[0, 0, 0, 0, 0, 0, 0, 0, 1, 0]。

```python
def parse_aug_fn(dataset):
    """
    Image Augmentation (图像增强) 函数
    """
    x = tf.cast(dataset['image'], tf.float32) / 255.  # 图像归一化
    x = flip(x)  # 随机水平翻转
    # 触发颜色转换概率 50%
    x = tf.cond(tf.random.uniform([], 0, 1) > 0.5, lambda:color(x), lambda: x)
    # 触发图像旋转概率 0.25%
    x = tf.cond(tf.random.uniform([], 0, 1) > 0.75, lambda:rotate(x), lambda: x)
    # 触发图像缩放概率 50%
    x = tf.cond(tf.random.uniform([], 0, 1) > 0.5, lambda: zoom(x), lambda: x)
    # 将输出标签转成独热编码
    y = tf.one_hot(dataset['label'], 10)
    return x, y

def parse_fn(dataset):
    x = tf.cast(dataset['image'], tf.float32) / 255.  # 图像归一化
    # 将输出标签转成独热编码
    y = tf.one_hot(dataset['label'], 10)
    return x, y
```

6.4.3 程序代码

Step 01 导入必要的套件。

```python
import os
import numpy as np
import pandas as pd
import tensorflow as tf
import tensorflow_datasets as tfds
import matplotlib.pyplot as plt
from tensorflow import keras
from tensorflow.keras import layers
from tensorflow.keras import initializers
# 从文件夹的 preprocessing.py 文件中 Import parse_aug_fn 和 parse_fn 函数
from preprocessing import parse_aug_fn, parse_fn
```

Step 02 读取数据并分析。

- 加载 CIFAR-10 数据集

```
# 将训练数据重新分成 1:9 等分，分别分给验证数据和训练数据
valid_split, train_split = tfds.Split.TRAIN.subsplit([10, 90])
# 获取训练数据，并读取数据的信息
train_data, info = tfds.load("cifar10", split=train_split, with_info=True)
# 获取验证数据
valid_data = tfds.load("cifar10", split=valid_split)
# 获取测试数据
test_data = tfds.load("cifar10", split=tfds.Split.TEST)
```

Step 03 数据集设置。

```
AUTOTUNE = tf.data.experimental.AUTOTUNE    # 自动调整模式
batch_size = 64    # 批量大小
train_num = int(info.splits['train'].num_examples / 10) * 9   # 训练数据数量

train_data = train_data.shuffle(train_num) # 打乱数据集
# 加载预处理 parse_aug_fn 函数，CPU 数量为自动调整模式
train_data = train_data.map(map_func=parse_aug_fn,
                            num_parallel_calls=AUTOTUNE)
# 设置批量大小，并启用 prefetch 模式（缓存空间为自动调整模式）
train_data = train_data.batch(batch_size).prefetch(buffer_size=AUTOTUNE)

# 加载预处理 parse_fn 函数，CPU 数量为自动调整模式
valid_data = valid_data.map(map_func=parse_fn, num_parallel_calls=AUTOTUNE)
# 设置批量大小，并启用 prefetch 模式（缓存空间为自动调整模式）
valid_data = valid_data.batch(batch_size).prefetch(buffer_size=AUTOTUNE)

# 加载预处理 parse_fn function，CPU 数量为自动调整模式
test_data = test_data.map(map_func=parse_fn, num_parallel_calls=AUTOTUNE)
# 设置批量大小，并启用 prefetch 模式（缓存空间为自动调整模式）
test_data = test_data.batch(batch_size).prefetch(buffer_size=AUTOTUNE)
```

Step 04 训练网络模型。

- 建立一个函数负责建立、编译和训练网络模型

    网络层配置如下：

    ➤ keras.Input：输入层（输入图像的大小为 32×32×3）。

    ➤ layers.Conv2D：卷积层（使用 ReLU 激活函数，以及 3×3 大小的 Kernel），这里 Kernel 是指卷积核。

    ➤ layers.MaxPool2D：池化层（对特征图下采样）。

    ➤ layers.Flatten：压平层（特征图转成一维张量）。

    ➤ layers.Dropout：随机失活层（每次训练随机失活 50% 的神经元）。

    ➤ layers.Dense：全连接层（隐藏层使用 ReLU 激活函数，输出层使用 Softmax 激活函数）。

```
def build_and_train_model(run_name, init):
    """
    run_name:传入目前执行的任务名字
    init:传入网络层初始化的方式
```

```python
    """
    inputs = keras.Input(shape=(32, 32, 3))
    x = layers.Conv2D(64, (3, 3), activation='relu',
                      kernel_initializer=init)(inputs)
    x = layers.MaxPool2D()(x)
    x = layers.Conv2D(128, (3, 3), activation='relu',
                      kernel_initializer=init)(x)
    x = layers.Conv2D(256, (3, 3), activation='relu',
                      kernel_initializer=init)(x)
    x = layers.Conv2D(128, (3, 3), activation='relu',
                      kernel_initializer=init)(x)
    x = layers.Conv2D(64, (3, 3), activation='relu', kernel_initializer=init)(x)
    x = layers.Flatten()(x)
    x = layers.Dense(64, activation='relu', kernel_initializer=init)(x)
    x = layers.Dropout(0.5)(x)
    outputs = layers.Dense(10, activation='softmax')(x)
    # 建立网络模型（将输入到输出所有经过的网络层连接起来）
    model = keras.Model(inputs, outputs)

    # 存储训练记录文件
    logfiles = 'lab5-logs/{}-{}'.format(run_name, init.__class__.__name__)
    # 存储各层权重分布
    model_cbk = keras.callbacks.TensorBoard(log_dir=logfiles,
                                            histogram_freq=1)
    # 存储最好的网络模型权重
    modelfiles = model_dir + '/{}-best-model.h5'.format(run_name)
    model_mckp = keras.callbacks.ModelCheckpoint(modelfiles,
                                monitor='val_categorical_accuracy',
                                save_best_only=True,
                                mode='max')
    # 设置训练使用的优化器、损失函数和评价指标函数
    model.compile(keras.optimizers.Adam(),
            loss=keras.losses.CategoricalCrossentropy(),
            metrics=[keras.metrics.CategoricalAccuracy()])

    # 训练网络模型
    model.fit(train_data,
            epochs=100,
            validation_data=valid_data,
            callbacks=[model_cbk, model_mckp])
```

- 训练 3 种权重初始化方法的网络模型

```python
session_num = 1
# 设置存储权重的目录
model_dir = 'lab5-logs/models/'
os.makedirs(model_dir)
# 设置要测试的 3 种权重初始化方法
weights_initialization_list = [initializers.RandomNormal(0, 0.01),
                               initializers.glorot_normal(),
```

```
                        initializers.he_normal()]
for init in weights_initialization_list:
    print('--- Running training session %d' % (session_num))
    run_name = "run-%d" % session_num
    build_and_train_model(run_name, init)  # 建立和训练网络
    session_num += 1
```

结果如图 6-16 所示。

```
Epoch 95/100
704/704 [==============================] - 13s 19ms/step - loss: 0.7215 - categorical_accuracy: 0.7616 - val_loss: 0.6090 - val_categorical_accuracy: 0.8026
Epoch 96/100
704/704 [==============================] - 13s 19ms/step - loss: 0.7239 - categorical_accuracy: 0.7618 - val_loss: 0.5869 - val_categorical_accuracy: 0.8136
Epoch 97/100
704/704 [==============================] - 13s 19ms/step - loss: 0.7152 - categorical_accuracy: 0.7641 - val_loss: 0.6050 - val_categorical_accuracy: 0.8096
Epoch 98/100
704/704 [==============================] - 13s 19ms/step - loss: 0.7254 - categorical_accuracy: 0.7603 - val_loss: 0.6007 - val_categorical_accuracy: 0.8082
Epoch 99/100
704/704 [==============================] - 14s 19ms/step - loss: 0.7207 - categorical_accuracy: 0.7644 - val_loss: 0.5784 - val_categorical_accuracy: 0.8144
Epoch 100/100
704/704 [==============================] - 14s 19ms/step - loss: 0.7297 - categorical_accuracy: 0.7622 - val_loss: 0.6065 - val_categorical_accuracy: 0.8162
```

图 6-16　执行结果

Step 05 比较 3 种初始化的训练结果。

- 读取各自最佳的网络权重

```
model_1 = keras.models.load_model('lab5-logs/models/run-1-best-model.h5')
model_2 = keras.models.load_model('lab5-logs/models/run-2-best-model.h5')
model_3 = keras.models.load_model('lab5-logs/models/run-3-best-model.h5')
```

- 在测试集数据上验证

```
loss_1, acc_1 = model_1.evaluate(test_data)
loss_2, acc_2 = model_2.evaluate(test_data)
loss_3, acc_3 = model_3.evaluate(test_data)
```

- 显示预测的损失值与正确率结果

从训练结果来看，He 初始化→Glorot 初始化→正态分布（std 0.01）初始化。

```
loss = [loss_1, loss_2, loss_3]
acc = [acc_1, acc_2, acc_3]
dict = {"Loss": loss, "Accuracy": acc}
pd.DataFrame(dict)
```

结果如图 6-17 所示。

	Loss	Accuracy
0	2.302650	0.1000
1	0.635572	0.7933
2	0.609081	0.8118

图 6-17　执行结果

6.4.4 TensorBoard 可视化权重分布

TensorBoard 提供了两种数据分布可视化的工具，分别为 DISTRIBUTIONS 和 HISTOGRAMS。前面章节训练网络时，tf.keras.callbacks.TensorBoard 回调函数把 histogram_freq 参数设置为 1，代表每训练一个 epoch，各层的网络权重分布就会被记录下来，并可以使用 DISTRIBUTIONS 和 HISTOGRAMS 两个可视化功能来查看权重分布的变化。

下面 TensorBoard 的操作示范将直接使用 6.4 节的训练结果，分别对正态分布（std 0.01）初始化、Glorot 初始化和 He 初始化 3 种初始化训练的权重分布变化进行分析。

 启动 TensorBoard（命令行）查看训练记录

```
tensorboard --logdir lab5-logs
```

 观察正态分布（std 0.01）初始化权重分布的变化

首先使用 DISTRIBUTIONS 工具来观察，图 6-18 显示标准差 0.01 正态分布初始化训练 100 个 epoch 的权重变化（x 轴为时间轴，y 轴为权重分布的范围）。从图中观察可知，第二层和第三层的卷积层的权重分布变动十分小，很明显网络权重完全没更新。

HISTOGRAMS 工具也可以用来观察分布的变化，图 6-19 以另一种形式显示正态分布（std 0.01）初始化训练 100 个 epoch 的权重变化（x 轴为权重分布的范围，y 轴为时间轴）。

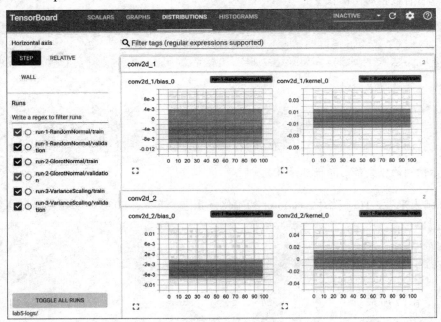

图 6-18　DISTRIBUTIONS 显示标准偏差 0.01 的正态分布初始化训练过程的权重变化
（图左上为第二层卷积层 Bias 分布，图右上为第二层卷积层 Kernel 分布，
图左下为第三层卷积层 Bias 分布，图右下为第三层卷积层 Kernel 分布）

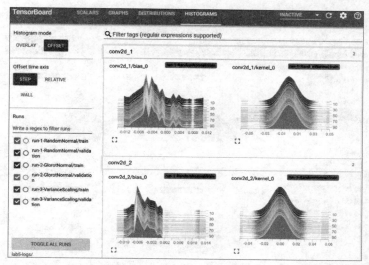

图 6-19　HISTOGRAMS 显示正态分布（std 0.01）初始化训练过程的权重变化
（图左上为第二层卷积层 Bias 分布，图右上为第二层卷积层 Kernel 分布，
图左下为第三层卷积层 Bias 分布，图右下为第三层卷积层 Kernel 分布）

观察 Glorot 初始化权重分布的变化

图 6-20 所示为 DISTRIBUTIONS 工具观察到的 Glorot 初始化第一层和第二层卷积层训练 100 个 epoch 的权重变化，相比正态分布（std 0.01）初始化的权重分布，Glorot 初始化训练越久，权重分布越广，代表能学习到更多样的特征。

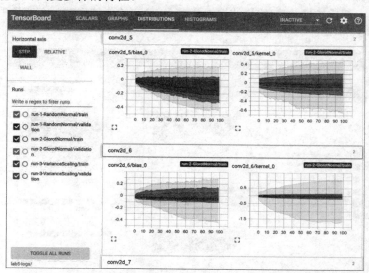

图 6-20　DISTRIBUTIONS 显示 Glorot 初始化训练过程的权重变化
（图左上为第一层卷积层 Bias 分布，图右上为第一层卷积层 Kernel 分布，
图左下为第二层卷积层 Bias 分布，图右下为第二层卷积层 Kernel 分布）

图 6-21 所示为 HISTOGRAMS 工具观察到的 Glorot 初始化第一层和第二层卷积层训练 100 个

epoch 的权重变化。

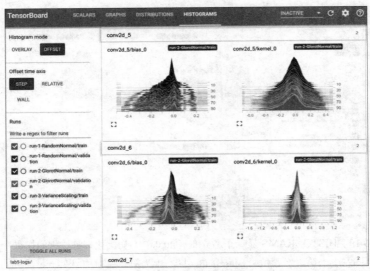

图 6-21　HISTOGRAMS 显示 Glorot 初始化训练过程的权重变化
（图左上为第一层卷积层 Bias 分布，图右上为第一层卷积层 Kernel 分布，
图左下为第二层卷积层 Bias 分布，图右下为第二层卷积层 Kernel 分布）

观察 He 初始化权重分布的变化

图 6-21 所示为 DISTRIBUTIONS 工具观察到的 He 初始化第一层和第二层卷积层训练 100 个 epoch 的权重变化，相比 Glorot 初始化的权重分布，He 初始化的权重分布感觉更分散且广，代表能学习到更多样的特征。

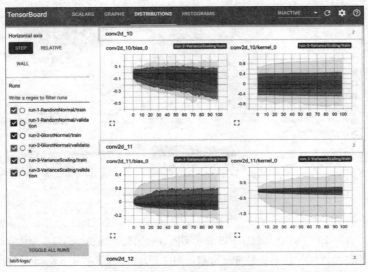

图 6-22　DISTRIBUTIONS 显示 He 初始化训练过程的权重变化
（图左上为第一层卷积层 Bias 分布，图右上为第一层卷积层 Kernel 分布，
图左下为第二层卷积层 Bias 分布，图右下为第二层卷积层 Kernel 分布）

图 6-23 所示为 HISTOGRAMS 工具观察到的 He 初始化第一层和第二层卷积层训练 100 个 epoch 的权重变化。

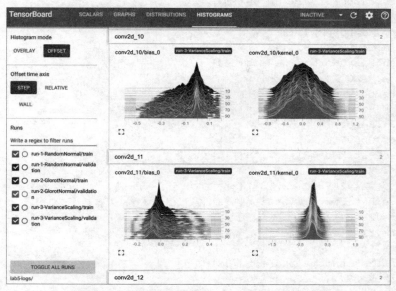

图 6-23 HISTOGRAMS 显示 He 初始化训练过程的权重变化
（图左上为第一层卷积层 Bias 分布，图右上为第一层卷积层 Kernel 分布，
图左下为第二层卷积层 Bias 分布，图右下为第二层卷积层 Kernel 分布）

6.5 实验二：使用 CIFAR-10 数据集实验批量归一化方法

本节继续采用 6.4 节的范例程序，在相同的网络架构上加入批量归一化，并用 CIFAR-10 数据集进行训练。

- 建立网络模型

这里使用了以下几种网络层：

> keras.Input：输入层（输入图像的大小为 32×32×3）。
> layers.Conv2D：卷积层（使用 3×3 大小的 Kernel），这里 Kernel 是指卷积核。
> layers.BatchNormalization：批量归一化层（使用默认参数）。
> layers.ReLU：ReLU 激活函数层（使用在 BatchNormalization 层之后）。
> layers.MaxPool2D：池化层（对特征图下采样）。
> layers.Flatten：压平层（特征图转成一维张量）。
> layers.Dropout：随机失活层（每次训练随机失活 50%的神经元）。
> layers.Dense：全连接层（隐藏层使用 ReLU 激活函数，输出层使用 Softmax 激活函数）。

```python
inputs = keras.Input(shape=(32, 32, 3))
x = layers.Conv2D(64, (3, 3))(inputs)
x = layers.BatchNormalization()(x)
x = layers.ReLU()(x)
x = layers.MaxPool2D()(x)
x = layers.Conv2D(128, (3, 3))(x)
x = layers.BatchNormalization()(x)
x = layers.ReLU()(x)
x = layers.Conv2D(256, (3, 3))(x)
x = layers.BatchNormalization()(x)
x = layers.ReLU()(x)
x = layers.Conv2D(128, (3, 3))(x)
x = layers.BatchNormalization()(x)
x = layers.ReLU()(x)
x = layers.Conv2D(64, (3, 3))(x)
x = layers.BatchNormalization()(x)
x = layers.ReLU()(x)
x = layers.Flatten()(x)
x = layers.Dense(64)(x)
x = layers.BatchNormalization()(x)
x = layers.ReLU()(x)
x = layers.Dropout(0.5)(x)
outputs = layers.Dense(10, activation='softmax')(x)

model_4 = keras.Model(inputs, outputs, name='model-4')
# 显示网络架构
model_4.summary()
```

结果如图 6-24 所示。

```
Model: "model-4"
_____
Layer (type)                 Output Shape              Param #
=================================================================
input_4 (InputLayer)         [(None, 32, 32, 3)]       0
_____
conv2d_15 (Conv2D)           (None, 30, 30, 64)        1792
_____
batch_normalization (BatchNo (None, 30, 30, 64)        256
_____
re_lu (ReLU)                 (None, 30, 30, 64)        0
_____
max_pooling2d_3 (MaxPooling2 (None, 15, 15, 64)        0
_____
conv2d_16 (Conv2D)           (None, 13, 13, 128)       73856
_____
batch_normalization_1 (Batch (None, 13, 13, 128)       512
_____
re_lu_1 (ReLU)               (None, 13, 13, 128)       0
_____
conv2d_17 (Conv2D)           (None, 11, 11, 256)       295168
_____
batch_normalization_2 (Batch (None, 11, 11, 256)       1024
_____
re_lu_2 (ReLU)               (None, 11, 11, 256)       0
_____
conv2d_18 (Conv2D)           (None, 9, 9, 128)         295040
_____
batch_normalization_3 (Batch (None, 9, 9, 128)         512
_____
re_lu_3 (ReLU)               (None, 9, 9, 128)         0
_____
conv2d_19 (Conv2D)           (None, 7, 7, 64)          73792
_____
batch_normalization_4 (Batch (None, 7, 7, 64)          256
_____
re_lu_4 (ReLU)               (None, 7, 7, 64)          0
_____
flatten_3 (Flatten)          (None, 3136)              0
_____
dense_6 (Dense)              (None, 64)                200768
_____
batch_normalization_5 (Batch (None, 64)                256
_____
re_lu_5 (ReLU)               (None, 64)                0
_____
dropout_3 (Dropout)          (None, 64)                0
_____
dense_7 (Dense)              (None, 10)                650
=================================================================
Total params: 943,882
Trainable params: 942,474
Non-trainable params: 1,408
```

图 6-24　执行结果

- 建立回调函数

```
model_dir = 'lab5-logs/models/'  # 设置存储权重的目录
# 存储训练记录文件
log_dir = os.path.join('lab5-logs', 'run-4-batchnormalization')
model_cbk = keras.callbacks.TensorBoard(log_dir=log_dir)
# 存储最好的网络模型权重
model_mckp = keras.callbacks.ModelCheckpoint(model_dir +
                            '/run-4-best-model.h5',
                            monitor='val_categorical_accuracy',
                            save_best_only=True,
                            mode='max')
```

- 设置训练使用的优化器、损失函数和评价指标函数

```
model_4.compile(keras.optimizers.Adam(),
          loss=keras.losses.CategoricalCrossentropy(),
          metrics=[keras.metrics.CategoricalAccuracy()])
```

- 训练网络模型

```
history_1 = model_4.fit(train_data,
                  epochs=100,
                  validation_data=valid_data,
                  callbacks=[model_cbk, model_mckp])
```

结果如图 6-25 所示。

```
Epoch 95/100
704/704 [==============================] - 17s 24ms/step - loss: 0.4179 - categorical_accuracy: 0.8642 - val_loss: 0.5299
 - val_categorical_accuracy: 0.8418
Epoch 96/100
704/704 [==============================] - 17s 24ms/step - loss: 0.4190 - categorical_accuracy: 0.8652 - val_loss: 0.8964
 - val_categorical_accuracy: 0.7376
Epoch 97/100
704/704 [==============================] - 17s 24ms/step - loss: 0.4236 - categorical_accuracy: 0.8633 - val_loss: 0.4678
 - val_categorical_accuracy: 0.8470
Epoch 98/100
704/704 [==============================] - 17s 25ms/step - loss: 0.4094 - categorical_accuracy: 0.8669 - val_loss: 0.4725
 - val_categorical_accuracy: 0.8548
Epoch 99/100
704/704 [==============================] - 17s 24ms/step - loss: 0.4174 - categorical_accuracy: 0.8640 - val_loss: 0.4594
 - val_categorical_accuracy: 0.8648
Epoch 100/100
704/704 [==============================] - 17s 24ms/step - loss: 0.4061 - categorical_accuracy: 0.8679 - val_loss: 0.4857
 - val_categorical_accuracy: 0.8552
```

图 6-25　执行结果

- 在测试集数据上验证

```
loss, acc = model_4.evaluate(test_data)
print('\nModel-4 Accuracy: {}%'.format(acc))
```

结果如下：

```
Model-4 Accuracy: 0.8593000173568726%
```

6.6 总结各种网络架构的性能比较

本节总结第 5、6 章所有网络模型训练的性能，表 6-2 所示为所有网络模型的差异性及性能比较表。

表 6-2 比较 7 种不同方法的性能

网络架构	IA	GU	RN	GN	HN	BN	Accuracy
全连接神经网络（Lab4 Model-1）	×	√	×	×	×	×	43.81%
卷积神经网络（Lab4 Model-2）	×	√	×	×	×	×	70.06%
卷积神经网络（Lab4 Model-3）	√	√	×	×	×	×	78.64%
卷积神经网络（Lab5 Model-1）	√	×	√	×	×	×	10.00%
卷积神经网络（Lab5 Model-2）	√	×	×	√	×	×	79.33%
卷积神经网络（Lab5 Model-3）	√	×	×	×	√	×	81.18%
卷积神经网络（Lab5 Model-4）	√	√	×	×	×	√	85.93%

※IA：图像增强（Image Augmentation）。
※GU：Glorot 均匀分布初始化（Glorot Uniform Initialization）。
※RN：随机正态分布初始化（Random Normal Distribution Initialization）。
※GN：Glorot 正态分布初始化（Glorot Normal Initialization）。
※HN：He 正态分布初始化（He Normal Initialization）。
※BN：批量归一化（Batch Normalization）。

也可以直接通过 TensorBoard（命令行）来观察第 5、6 章的训练记录，如图 6-26 所示。

```
tensorboard --logdir lab4-logs/,lab5-logs
```

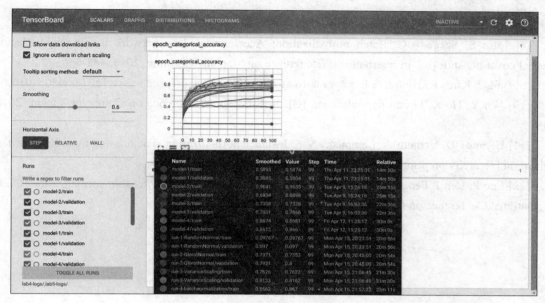

图 6-26　TensorBoard 显示结果

> **注　意**
>
> --logdir 后需要输入两个记录文件目录所在的位置。

6.7　参考文献

[1] Rumelhart D E, Hinton G E, Williams R J. Learning representations by back-propagation errors [J]. Nature, 1986, 533-536.

[2] Mishkin D, Matas J. All you need is a good init[C]. in Proc. International Conference on Learning Representations, 2016, 3013-3018.

[3] Xie D, Xiong J, Pu S. All you need is beyond a good init: Exploring better solution for training extremely deep convolutional neural networks with orthonormality and modulation [J]. Proceedings of the IEEE Conference on Computer Vision and Pattern Recognition, 2017, 6176-6185.

[4] Salimans T, Kingma D P. Weight normalization: A simple reparameterization to accelerate training of deep neural networks[C]. Advances in Neural Information Processing Systems, 2016, 901-909.

[5] Masood S, Doja M N, Chandra P. Analysis of weight initialization methods for gradient descent with momentum [C]. In International Conference on Soft Computing Techniques and Implementations, 2015, 131-136.

[6] Glorot X, Bengio Y. Understanding the difficulty of training deep feedforward neural networks [C]. In Proc. Conf. Artificial Intelligence and Statistics, 2010, 249-256.

[7] He K, Zhang X, Ren S, et al. Delving Deep into Rectifiers: Surpassing Human-Level Performance on ImageNet Classification [C]. In IEEE International Conference on Computer Vision,

2015, 1026-1034.

[8] Ioffe S, Szegedy C. Batch normalization: Accelerating deep network training by reducing internal covariate shift [C]. In International Conference on Machine Learning, 2015, 448-456.

[9] Ba L J, Kiros R, Hinton G E. Layer normalization [J]. arXiv preprint arXiv:1607.06450, 2016.

[10] Wu Y, He K. Group normalization [C]. In European Conference on Computer Vision, 2018, 3-19.

[11] Ulyanov D, Vedaldi A, Lempitsky V S. Instance normalization: The missing ingredient for fast stylization [J]. arXiv preprint arXiv:1607.08022, 2016.

[12] Luo P, Ren J, Peng Z. Differentiable learning-to-normalize via switchable normalization [C]. In International Conference on Learning Representations, 2019.

第 7 章

TensorFlow 2.0 高级技巧

学习目标

- 更深入地使用 TensorFlow 2.0，建立自定义的 API，例如自定义网络层、自定义损失函数、自定义评价指标函数和自定义回调函数
- 利用自定义 API 来训练网络模型，并且训练结果与 Keras 高级 API 相近

7.1　TensorFlow 高级技巧

前面几个章节介绍了如何使用 tf.keras 搭建网络模型和训练网络模型，并且使用少量的程序代码完成了 3 个深度学习入门任务：回归任务、二分类任务和多分类任务。每个项目都少不了以下几个步骤：

（1）搭建网络内部层：tf.keras.layers。
（2）建立网络模型：tf.keras.Model。
（3）设置训练优化器、损失函数及评价指标函数：model.compile。
（4）训练网络模型：model.fit。

尽管 TensorFlow 的 Keras 高级 API 让搭建和训练网络模型变得十分简单，但是官方提供的高级 API 还是有局限性，并不是所有新的网络架构或损失函数都会包含在高级 API 当中。例如，在实现新论文的网络架构时，作者提出的网络层或损失函数在这些 API 中都找不到。所以，本章将介绍 TensorFlow 的高级技巧，教导读者如何定义自己的 API，内容如下：

- 自定义网络层（Layer）：介绍如何建立全新的网络层。
- 自定义损失函数（Loss）：介绍如何建立全新的损失函数。

- 自定义评价指标函数（Metrics）：介绍如何建立全新的评价指标函数。
- 自定义回调函数（Callback）：介绍如何建立全新的回调函数。

7.1.1 自定义网络层

TensorFlow 提供的网络层种类相当多，分别说明如下：

- 卷积层：Conv1D、Conv2D、Conv3D、SeparableConv2D、DepthwiseConv2D、Conv2DTranspose 等。
- 池化层：MaxPooling1D、MaxPooling2D、AveragePooling2D、GlobalMaxPooling2D 等。
- 融合层：Add、Subtract、Multiply、Concatenate、Maximum 等。
- 激活层：ReLU、LeakyReLU、PReLU 等。
- 循环层：RNN、GRU、LSTM、ConvLSTM2D 等。
- 批量归一化层、随机失活层等。

若没有找到我们想要的网络层，则可通过继承 tf.keras.layers.Layer 类来建立自定义的网络层。

> **提 示**
>
> TensorFlow 官方文件可参考：
> https://www.tensorflow.org/versions/r2.0/api_docs/python/tf/keras/layers

下面是继承 Layer 类的方法：

```
class CustomLayer(tf.keras.layers.Layer):
    def __init__(self, **kwargs):
        super(CustomLayer, self).__init__(**kwargs)
        """
        设置参数的地方。
        """
    def build(self, input_shape):
        """
        建立权重的地方（通过 add_weight 方法）。
        参数：
            input_shape:输入大小。
        """
    def call(self, inputs):
        """
        定义网络前向传播（运算）的地方。
        参数：
            inputs:输入网络的数据。
        """
    def get_config(self):
        """
        （选择）如果要支持序列化，就在这里定义，它会返回层的构建参数。
        """
```

7.1.2 自定义损失函数

设计新的架构或实现新论文中的架构时，TensorFlow 官方文件所提供的损失函数，通常不足以应付所有问题，此时我们就必须自己定义损失函数。

> **提 示**
>
> TensorFlow 官方文件可参考：
> https://www.tensorflow.org/versions/r2.0/api_docs/python/tf/keras/losses

下面是自定义损失函数的模板：

```
def custom_loss(y_true, y_pred):
    """
    定义 loss 计算在这个地方。
    参数：
        y_true(真实值)：传入这笔数据的答案。
        y_pred(预测值)：传入这笔数据网络预测的结果。
    """
    return loss
```

7.1.3 自定义评价指标函数

评价指标函数用来评估模型的好坏。深度学习领域本身有非常多种类的任务，例如对象检测（Object Detection）算法，要计算 MAP（Mean Average Precision，平均精度均值）来评估模型的好坏，而 TensorFlow 就没有提供类似的评价指标函数，此时我们可以通过继承 tf.keras.metrics.Metric 类来建立自定义的评价指标函数。

> **提 示**
>
> TensorFlow 官方文件可参考：
> https://www.tensorflow.org/versions/r2.0/api_docs/python/tf/keras/metrics

下面是继承 Metrics 类的方法：

```
class CustomMetrics(tf.keras.metrics.Metric):
    def __init__(self, name='custom_metrics', **kwargs):
        super(CustomMetrics, self).__init__(name=name, **kwargs)
        """
        所有评价指标函数使用到的状态变量都需在这里建立。
        参数：
            name：评价指标函数的名称。
        """
    def update_state(self, y_true, y_pred, sample_weight=None):
        """
        使用 y_true(真实值)与 y_pred(预测值)来计算更新状态变量。
```

```
        参数:
            y_true(真实值):传入这笔数据的答案。
            y_pred(预测值):传入这笔数据网络预测的结果。
            sample_weight:对样本的权重,通常用于序贯模型。
        """
    def result(self):
        """
        使用状态变量计算最终的结果。
        """
    def reset_states(self):
        """
        重新初始化评价指标函数(状态变量)。
        """
```

7.1.4 自定义回调函数

回调函数主要是在执行 model.fit、model.evaluate 或 model.predict 等过程中调用,因为这些操作每次执行都可能花费很长的时间,所以需要通过回调函数从中监控、记录或调整模型。简单来说,网络在训练过程中无法干涉,需要通过回调函数代为转达或执行。

下面列出几个常用的回调函数:

- tf.keras.callbacks.ModelCheckpoint:监测数值,将最好的模型权重存储下来。
- tf.keras.callbacks.EarlyStopping:如果模型的监测数值太久没有进步,就会提前终止训练。
- tf.keras.callbacks.ReduceLROnPlateau:如果模型的监测数值太久没有进步,就将会降低学习率。
- tf.keras.callbacks.TensorBoard:记录模型训练过程、权重和图。

如果所需的监控或执行的功能在官方的回调函数中没有提供,那么可以通过继承 tf.keras.callbacks.Callback 类来建立自定义的回调函数。

> **提 示**
>
> TensorFlow 官方文件可参考:
> https://www.tensorflow.org/versions/r2.0/api_docs/python/tf/keras/callbacks

下面是继承 Callback 类的方法:

```
class CustomCallbacks(tf.keras.callbacks.Callback):
    def on_epoch_(begin|end)(self, epoch, logs=None):
        """
        每一个 epoch 开始或结束,执行这段程序。
        参数:
            epoch:目前的 epoch。
            logs:传入 dict 格式的记录信息,例如 loss、val_loss 等。
        """
    def on_(train|test|predict)_begin(self, logs=None):
        """
```

```
        fit、evaluate 或 predict 任务开始时，执行这段程序。
        参数：
            logs：传入 dict 格式的记录信息，例如 loss、val_loss 等。
        """
    def on_(train|test|predict)_end(self, logs=None):
        """
        fit、evaluate 或 predict 任务结束时，执行这段程序。
        参数：
            logs：传入 dict 格式的记录信息，例如 loss、val_loss 等。
        """
    def on_(train|test|predict)_batch_begin(self, batch, logs=None):
        """
        fit、evaluate 或 predict 任务的每一个批量 batch 开始前执行这段程序。
        参数：
            batch：目前的 batch。
            logs：传入 dict 格式的记录信息，例如 loss、val_loss 等。
        """
    def on_(train|test|predict)_batch_end(self, batch, logs=None):
        """
        fit、evaluate 或 predict 任务的每一个批量 batch 结束后执行这段程序。
        参数：
            batch：目前的 batch。
            logs：传入 dict 格式的记录信息，例如 loss、val_loss 等。
        """
```

7.2 Keras 高级 API 与自定义 API 比较

介绍完 TensorFlow 高级技巧后，我们已经初步了解了自定义 API。接下来，本节会以实际的范例程序来加深读者印象，每个范例程序都会提供两种做法：Keras 高级 API 和自定义 API，这两种方法能够实现相同的功能。

7.2.1 网络层

本节介绍两种建立卷积层的方法，第一种通过 Keras 高级 API，第二种通过自定义网络层，卷积层的参数如下：

- Kernel 数量：64。
- Kernel 大小：3×3。
- Strides：1。
- Padding：valid。
- 激活函数：ReLU。
- Kernel 初始化：glorot_uniform。
- Bias 初始化：zeros。

下面两种方法建立的卷积层具有相同的功能：

- 方法一：使用高级的 TensorFlow API 来建立卷积层。

```
tf.keras.layers.Conv2D(64, 3, activation='relu',
                kernel_initializer='glorot_uniform')
```

- 方法二：使用自定义网络层来建立卷积层。

```
class CustomConv2D(tf.keras.layers.Layer):
    def __init__(self, filters, kernel_size, strides=(1, 1), padding="VALID",
            **kwargs):
        super(CustomConv2D, self).__init__(**kwargs)
        self.filters = filters
        self.kernel_size = kernel_size
        self.strides = (1, *strides, 1)
        self.padding = padding

    def build(self, input_shape):
        kernel_h, kernel_w = self.kernel_size
        input_dim = input_shape[-1]
        # 建立卷积层的权重值(weights)
        self.w = self.add_weight(name='kernel',
                    shape=(kernel_h, kernel_w, input_dim, self.filters),
                    initializer='glorot_uniform',   # 设置初始化方法
                    trainable=True)         # 设置这个权重是否能够训练
        # 建立卷积层的偏差值（bias）
        self.b = self.add_weight(name='bias',
                    shape=(self.filters,),
                    initializer='zeros',       # 设置初始化方法
                    trainable=True)         # 设置这个权重是否能够训练

    def call(self, inputs):
        # 卷积运算
        x = tf.nn.conv2d(inputs, self.w, self.strides, padding=self.padding)
        x = tf.nn.bias_add(x, self.b)          # 加上偏差值
        x = tf.nn.relu(x)                # 激活函数
        return x
```

7.2.2 损失函数

本节介绍两种建立多分类交叉熵的方法，第一种通过 Keras 高级 API，第二种通过自定义损失函数，损失函数公式如下：

$$CCE = -\frac{\sum_{i=1}^{N}\sum_{j=0}^{C} y_{i,j} \log(f(\hat{y}_{i,j}))}{N}$$

y：预期输出值。
$\hat{y}$：深度学习模型没经过 Softmax 的输出值。

f：Softmax 函数。
C：类别数量。
N：一个批量的数据量。

下面两种方法建立的损失函数具有相同的功能：

- 方法一：使用高级的 TensorFlow API 来建立多分类交叉熵。

```
tf.keras.losses.CategoricalCrossentropy()
```

- 方法二：使用自定义损失函数来建立多分类交叉熵。

```
def custom_categorical_crossentropy(y_true, y_pred):
    x = tf.reduce_mean(-tf.reduce_sum(y_true * tf.log(y_pred),
                reduction_indices=[1]))
    return x
```

实际训练时并不会使用上面自定义的交叉熵损失函数，而是使用官方提供的 tf.nn.softmax_cross_entropy_with_logits 低级 API 来实现，主要是因为官方提供的交叉熵 API 经过优化，在训练上速度更快且更加稳定。

```
def custom_categorical_crossentropy(y_true, y_pred):
    x = tf.nn.softmax_cross_entropy_with_logits(labels=y_true, logits=y_pred)
    return x
```

7.2.3 评价指标函数

本小节介绍两种建立分类正确率（Categorical Accuracy）的方法，第一种通过 Keras 高级 API，第二种通过自定义评价指标函数。分类正确率是用来计算多分类问题的正确率的指标，计算公式如下：

$$\text{Categorical Accuracy} = \frac{\text{Correct numbers}}{\text{Total numbers}}$$

Correct numbers：正确预测的数量。
Total numbers：全部数据的数量。

下面两种方法建立的评价指标函数具有相同的功能：

- 方法一：使用高级的 TensorFlow API 建立分类正确率。

```
tf.keras.metrics.CategoricalAccuracy()
```

- 方法二：使用自定义评价指标函数建立分类正确率。

```
class CustomCategoricalAccuracy(tf.keras.metrics.Metric):
    def __init__(self, name='custom_catrgorical_accuracy', **kwargs):
        super(CustomCategoricalAccuracy, self).__init__(name=name, **kwargs)
        # 记录正确预测的数量
        self.correct = self.add_weight('correct_numbers',
                                initializer='zeros')
```

```python
        # 记录全部数据的数量
        self.total = self.add_weight(total_numbers, initializer='zeros')

    def update_state(self, y_true, y_pred, sample_weight=None):
        # 输入答案为独热编码，所以取最大的数值为答案
        y_true = tf.argmax(y_true, axis=-1)
        # 取预测输出最大的数值为预测结果
        y_pred = tf.argmax(y_pred, axis=-1)
        # 比较预测结果是否正确，正确会返回 True，错误会返回 False
        values = tf.equal(y_true, y_pred)
        # 转换为浮点数，True=1.0, False=0.0
        values = tf.cast(values, tf.float32)
        # 将 values 所有数值相加就会等于正确预测的总数
        values_sum = tf.reduce_sum(values)
        # 计算这个批量的数据数量
        num_values = tf.size(values, out_type=tf.float32)
        self.correct.assign_add(values_sum)      # 更新正确预测的总数
        self.total.assign_add(num_values)        # 更新数据量的总数

    def result(self):
        # 计算正确率
        return tf.math.divide_no_nan(self.correct, self.total)

    def reset_states(self):
        # 每一次 epoch 结束后会重新初始化变量
        self.correct.assign(0.)
        self.total.assign(0.)
```

> **说 明**
>
> **分类正确率运算范例**
>
> 输入 3 笔数据进入网络，网络的预测输出为 y_pred，3 笔数据对应的答案为 y_true（用独热编码表示）。下面的程序是 3 笔数据预测的正确率的计算。
>
> ```
> # y_true(答案为独热编码)为二维张量, Tensor shape=(3, 10)
> y_true = tf.constant([[0, 0, 0, 1, 0, 0, 0, 0, 0],
>                       [0, 0, 0, 0, 1, 0, 0, 0, 0],
>                       [1, 0, 0, 0, 0, 0, 0, 0, 0]])
> # y_pred(预测输出值)为二维张量, Tensor shape=(3, 10)
> y_pred = tf.constant( [[.1, 0, 0, .7, 0, 0, 0, 0.2, 0],
>                        [0, 0, 0, 0, .8, .1, 0, 0, .1],
>                        [.3, 0, 0, 0, 0, 0, 0, .1, .6]])
> # 取 y_true 最大的数值为答案, axis=-1代表在张量的最后一维进行运算
> # tf.argmax([[0, 0, 0, 1, 0, 0, 0, 0, 0], [...], [...]], axis=-1) → [3, 4, 0]
> y_true = tf.argmax(y_true, axis=-1)
> # 取 y_pred 最大的数值为预测结果, axis=-1 代表在张量的最后一维进行运算
> # tf.argmax([[.1, 0, 0, .7, 0, 0, 0, 0.2, 0], [...], [...]], axis=-1) → [3, 4, 8]
> y_pred = tf.argmax(y_pred, axis=-1)
> # 比较预测结果是否正确，正确则返回 True，错误则返回 False
> ```

> 说明（续）
> ```
> # tf.equal([3, 4, 0], [3, 4, 8])→ [True, True, False] values = tf.equal(y_true, y_pred)
> # 转换为浮点数 (True=1.0, False=0.0)
> # tf.cast([True, True, False], tf.float32)→ [1. , 1. , 0.]
> values = tf.cast(values, tf.float32)
> # 将一个批量的数据加总
> # tf.reduce_sum([1. , 1. , 0.])→ 2.0 values_sum = tf.reduce_sum(values)
> # 计算这个批量的数据数量
> # tf.size([1. 1. 0.], out_type=tf.float32)→ 3.0
> num_values = tf.cast(tf.size(values), tf.float32)
> ```

7.2.4 回调函数

本小节介绍两种建立存储模型权重的方法，第一种通过 Keras 高级 API，第二种通过自定义回调函数。存储模型权重的方法会存储性能最好的模型权重，如图 7-1 所示。

图 7-1 存储模型权重

下面两种方法建立的回调函数具有相同的功能：

- 方法一：使用高级的 TensorFlow API 建立存储模型权重回调函数。

```
tf.keras.callbacks.ModelCheckpoint('logs/models/save.h5')
```

- 方法二：使用自定义回调函数建立存储模型权重回调函数。

```python
class SaveModel(tf.keras.callbacks.Callback):
    def __init__(self, weights_file, monitor='loss', mode='min',
                 save_weights_only=False):
        super(SaveModel, self).__init__()
        # 设置模型权重的存储路径
        self.weights_file = weights_file
        # 设置要监测的数值
        self.monitor = monitor
        # 设置监测值越大越好还是越小越好
        # Ex:如果监测数值为loss时，就设为'min'；如果监测数值为Accuracy，就设为'max'
```

```python
        self.mode = mode
        # 只存储网络权重或存储整个网络模型（包含 Layer、Compile 等）
        self.save_weights_only = save_weights_only
        if mode == 'min':
            # 设置 best 为无限大
            self.best = np.Inf
        else:
            # 设置 best 为负无限大
            self.best = -np.Inf

    # 存储网络模型的函数
    def save_model(self):
        if self.save_weights_only:
            # 只存储网络权重
            self.model.save_weights(self.weights_file)
        else:
            # 存储整个网络模型（包含 Layer、Compile 等）
            self.model.save(self.weights_file)

    def on_epoch_end(self, epoch, logs=None):
        # 从 logs 中读取监测值
        monitor_value = logs.get(self.monitor)
        # 如果监测值降低或升高(取决于 mode 设置)，就存储网络模型
        if self.mode == 'min' and monitor_value < self.best:
            self.save_model()
            self.best = monitor_value
        elif self.mode == 'max' and monitor_value > self.best:
            self.save_model()
            self.best = monitor_value
```

7.3 实验：比较 Keras 高级 API 和自定义 API 两种网络训练的结果

本实验为了验证 Keras 高级 API 与自定义 API 两种方法拥有相同的功能，所以在 CIFAR-10 的范例程序中分别使用这两种方法训练网络模型，希望自定义 API 训练出来的网络模型性能可以和 Keras 高级 API 相近。最后实验结果显示两种方法训练出来的网络模型正确率非常接近，因此证实自定义 API 也能达到和 Keras 高级 API 相同的功能。

7.3.1 新建项目

建议使用 Jupyter Notebook 来执行本章的程序代码，操作流程如下：

Step 01 启动 Jupyter Notebook。

在 Terminal（Ubuntu）或命令提示符（Windows）中输入如下指令：

```
jupyter notebook
```

Step 02 新建执行文件。

单击界面右上角的 New 下拉按钮，然后单击所安装的 Python 解释器（在 Jupyter 中都称为 Kernel）来启动它，如图 7-2 所示，显示了 3 个不同的 Kernel：

- Python 3：本地端 Python。
- tf2：虚拟机 Python（TensorFlow-cpu 版本）。
- tf2-gpu：虚拟机 Python（TensorFlow-gpu 版本）。

图 7-2　新建执行文件

Step 03 执行程序代码。

按 Shift + Enter 快捷键执行单行程序代码，如图7-3 所示。

图 7-3　Jupyter 环境界面

接下来，后续的程序代码都可以在 Jupyter Notebook 上执行。

7.3.2　程序代码

Step 01 导入必要的套件。

```
import os
import numpy as np
import pandas as pd
import tensorflow as tf
```

```python
import tensorflow_datasets as tfds
from tensorflow import keras
from tensorflow.keras import layers
from tensorflow.keras import initializers
# 从文件夹的 preprocessing.py 文件中 Import parse_aug_fn 和 parse_fn 函数
from preprocessing import parse_aug_fn, parse_fn
```

Step 02 读取数据。

- 加载 CIFAR-10 数据集

```python
# 将训练数据重新分成 1:9 等分，分别分给验证数据和训练数据
valid_split, train_split = tfds.Split.TRAIN.subsplit([10, 90])
# 获取训练数据，并读取 data 的信息
train_data, info = tfds.load("cifar10", split=train_split, with_info=True)
# 获取验证数据
valid_data = tfds.load("cifar10", split=valid_split)
# 获取测试数据
test_data = tfds.load("cifar10", split=tfds.Split.TEST)
# 获取 CIFAR-10 数据集的类别
class_name = info.features['label'].names
```

Step 03 数据集设置。

```python
AUTOTUNE = tf.data.experimental.AUTOTUNE      # 自动调整模式
batch_size = 64  # 批量大小
train_num = int(info.splits['train'].num_examples / 10) * 9  # 训练数据数量

train_data = train_data.shuffle(train_num) # 打乱数据集
# 加载预处理 parse_aug_fn 函数，CPU 数量为自动调整模式
train_data = train_data.map(map_func=parse_aug_fn, num_parallel_calls=AUTOTUNE)
# 设置批量大小，并启用 prefetch 模式（缓存空间为自动调整模式）
train_data = train_data.batch(batch_size).prefetch(buffer_size=AUTOTUNE)

# 加载预处理 parse_fn 函数，CPU 数量为自动调整模式
valid_data = valid_data.map(map_func=parse_fn, num_parallel_calls=AUTOTUNE)
# 设置批量大小，并启用 prefetch 模式（缓存空间为自动调整模式）
valid_data = valid_data.batch(batch_size).prefetch(buffer_size=AUTOTUNE)

# 加载预处理 parse_fn 函数，CPU 数量为自动调整模式
test_data = test_data.map(map_func=parse_fn, num_parallel_calls=AUTOTUNE)
# 设置批量大小，并启用 prefetch 模式（缓存空间为自动调整模式）
test_data = test_data.batch(batch_size).prefetch(buffer_size=AUTOTUNE)
```

Step 04 使用 Keras 高级 API 训练网络模型。

- 建立网络模型

这里使用了以下几种网络层：

➢ keras.Input：输入层（输入图像的大小为 32×32×3）。

- layers.Conv2D：卷积层（使用 ReLU 激活函数，以及 3×3 大小的 Kernel），这里 Kernel 是指卷积核。
- layers.MaxPool2D：池化层（对特征图下采样）。
- layers.Flatten：压平层（特征图转成一维张量）。
- layers.Dropout：随机失活层（每次训练随机失活 50% 的神经元）。
- layers.Dense：全连接层（隐藏层使用 ReLU 激活函数，最后一层使用线性输出）。

```
inputs = keras.Input(shape=(32, 32, 3))
x = layers.Conv2D(64, 3, activation='relu',
                  kernel_initializer='glorot_uniform')(inputs)
x = layers.MaxPool2D()(x)
x = layers.Conv2D(128, 3, activation='relu',
                  kernel_initializer='glorot_uniform')(x)
x = layers.Conv2D(256, 3, activation='relu',
                  kernel_initializer='glorot_uniform')(x)
x = layers.Conv2D(128, 3, activation='relu',
                  kernel_initializer='glorot_uniform')(x)
x = layers.Conv2D(64, 3, activation='relu',
                  kernel_initializer='glorot_uniform')(x)
x = layers.Flatten()(x)
x = layers.Dense(64, activation='relu')(x)
x = layers.Dropout(0.5)(x)
outputs = layers.Dense(10)(x)
# 建立网络模型（将输入到输出所有经过的网络层连接起来）
model_1 = keras.Model(inputs, outputs, name='model 1')
model_1.summary()
```

结果如图 7-4 所示。

```
Model: "model 1"
_____
Layer (type)                 Output Shape              Param #
=================================================================
input_5 (InputLayer)         [(None, 32, 32, 3)]       0
conv2d_7 (Conv2D)            (None, 30, 30, 64)        1792
max_pooling2d_4 (MaxPooling2 (None, 15, 15, 64)        0
conv2d_8 (Conv2D)            (None, 13, 13, 128)       73856
conv2d_9 (Conv2D)            (None, 11, 11, 256)       295168
conv2d_10 (Conv2D)           (None, 9, 9, 128)         295040
conv2d_11 (Conv2D)           (None, 7, 7, 64)          73792
flatten_4 (Flatten)          (None, 3136)              0
dense_8 (Dense)              (None, 64)                200768
dropout_4 (Dropout)          (None, 64)                0
dense_9 (Dense)              (None, 10)                650
=================================================================
Total params: 941,066
Trainable params: 941,066
Non-trainable params: 0
```

图 7-4　执行结果

- 建立回调函数

```
# 存储训练记录文件
logs_dirs = 'lab6-logs'
model_cbk = keras.callbacks.TensorBoard(log_dir='lab6-logs')
# 创建存储模型权重的目录
model_dirs = logs_dirs + '/models'
os.makedirs(model_dirs, exist_ok=True)
save_model = keras.callbacks.ModelCheckpoint(model_dirs + '/save.h5',
                                             monitor='val_catrgorical_accuracy',
                                             mode='max')
```

- 设置训练使用的优化器、损失函数和评价指标函数

```
model_1.compile(keras.optimizers.Adam(),
                # 由于网络输出没有 Softmax，因此要将 from_logits 设为 True
                loss=keras.losses.CategoricalCrossentropy(from_logits=True),
                metrics=[keras.metrics.CategoricalAccuracy()])
```

- 训练网络模型

```
# 训练网络模型
model_1.fit(train_data,
            epochs=100,
            validation_data=valid_data,
            callbacks=[model_cbk, save_model])
```

结果如图 7-5 所示。

```
Epoch 96/100
704/704 [==============================] - 12s 17ms/step - loss: 0.8100 - categorical_accuracy: 0.7326 - val_loss: 0.6121
- val_categorical_accuracy: 0.7950
Epoch 97/100

704/704 [==============================] - 12s 17ms/step - loss: 0.8067 - categorical_accuracy: 0.7320 - val_loss: 0.6470
- val_categorical_accuracy: 0.7874
Epoch 98/100
704/704 [==============================] - 12s 17ms/step - loss: 0.7964 - categorical_accuracy: 0.7348 - val_loss: 0.6376
- val_categorical_accuracy: 0.7856
Epoch 99/100
704/704 [==============================] - 12s 17ms/step - loss: 0.7924 - categorical_accuracy: 0.7360 - val_loss: 0.6556
- val_categorical_accuracy: 0.7892
Epoch 100/100
704/704 [==============================] - 12s 17ms/step - loss: 0.7985 - categorical_accuracy: 0.7351 - val_loss: 0.6480
- val_categorical_accuracy: 0.7882

<tensorflow.python.keras.callbacks.History at 0x7f8288443eb8>
```

图 7-5 执行结果

Step 05 使用自定义 API 训练网络模型。

- 建立 CustomConv2D 卷积层

```
class CustomConv2D(tf.keras.layers.Layer):
    def __init__(self, filters, kernel_size, strides=(1, 1), padding="VALID",
                 **kwargs):
        super(CustomConv2D, self).__init__(**kwargs)
        self.filters = filters
        self.kernel_size = kernel_size
        self.strides = (1, *strides, 1)
        self.padding = padding
```

```python
    def build(self, input_shape):
        kernel_h, kernel_w = self.kernel_size
        input_dim = input_shape[-1]
        # 建立卷积层的权重值(weights)
        self.w = self.add_weight(name='kernel',
                            shape=(kernel_h, kernel_w, input_dim, self.filters),
                            initializer='glorot_uniform',  # 设置初始化方法
                            trainable=True)  # 设置这个权重是否能够训练
        # 建立卷积层的偏差值(bias)
        self.b = self.add_weight(name='bias',
                            shape=(self.filters,),
                            initializer='zeros',  # 设置初始化方法
                            trainable=True)  # 设置这个权重是否能够训练

    def call(self, inputs):
        # 卷积运算
        x = tf.nn.conv2d(inputs, self.w, self.strides, padding=self.padding)
        x = tf.nn.bias_add(x, self.b)  # 加上偏差值
        x = tf.nn.relu(x)  # 激活函数
        return x
```

- 建立网络模型

这里使用了以下几种网络层，其中将 layers.Conv2D 换成自定义卷积层，其余架构则保留不变：

> keras.Input：输入层（输入图像的大小为 32×32×3）。
> CustomConv2D：自定义卷积层。
> layers.MaxPool2D：池化层（对特征图下采样）。
> layers.Flatten：压平层（特征图转成一维张量）。
> layers.Dropout：随机失活层（每次训练随机失活 50%的神经元）。
> layers.Dense：全连接层（隐藏层使用 ReLU 激活函数，最后一层使用线性输出）。

```python
inputs = keras.Input(shape=(32, 32, 3))
x = CustomConv2D(64, (3, 3))(inputs)       # 自定义卷积层
x = layers.MaxPool2D()(x)
x = CustomConv2D(128, (3, 3))(x)           # 自定义卷积层
x = CustomConv2D(256, (3, 3))(x)           # 自定义卷积层
x = CustomConv2D(128, (3, 3))(x)           # 自定义卷积层
x = CustomConv2D(64, (3, 3))(x)            # 自定义卷积层
x = layers.Flatten()(x)
x = layers.Dense(64, activation='relu')(x)
x = layers.Dropout(0.5)(x)
outputs = layers.Dense(10)(x)
# 建立网络模型（将输入到输出所有经过的网络层连接起来）
model_2 = keras.Model(inputs, outputs, name='model 2')
model_2.summary()
```

结果如图 7-6 所示。

```
Model: "model_2"
_____
Layer (type)                 Output Shape              Param #
=================================================================
input_6 (InputLayer)         [(None, 32, 32, 3)]       0
_____
custom_conv2d_15 (CustomConv (None, 30, 30, 64)        1792
_____
max_pooling2d_5 (MaxPooling2 (None, 15, 15, 64)        0
_____
custom_conv2d_16 (CustomConv (None, 13, 13, 128)       73856
_____
custom_conv2d_17 (CustomConv (None, 11, 11, 256)       295168
_____
custom_conv2d_18 (CustomConv (None, 9, 9, 128)         295040
_____
custom_conv2d_19 (CustomConv (None, 7, 7, 64)          73792
_____
flatten_5 (Flatten)          (None, 3136)              0
_____
dense_10 (Dense)             (None, 64)                200768
_____
dropout_5 (Dropout)          (None, 64)                0
_____
dense_11 (Dense)             (None, 10)                650
=================================================================
Total params: 941,066
Trainable params: 941,066
Non-trainable params: 0
```

图 7-6　执行结果

> **注　意**
>
> 最后一层输出不要加上 Softmax 激活函数，因为自定义的损失函数使用 tf.nn.softmax_cross_entropy_with_logits API 来计算损失，而这个 API 会将输入的数值经过一次 Softmax 再计算交叉熵，因此最后一层若使用 Softmax 激活函数，则输出数值会经过两次 Softmax 再计算交叉熵。

- 建立 SaveModel 回调函数

```python
class SaveModel(tf.keras.callbacks.Callback):
    def __init__(self, weights_file, monitor='loss', mode='min',
                 save_weights_only=False):
        super(SaveModel, self).__init__()
        # 设置模型权重的存储路径
        self.weights_file = weights_file
        # 设置要监测的数值
        self.monitor = monitor
        # 设置监测值越大越好还是越小越好
        self.mode = mode
        # 只存储网络权重或存储整个网络模型(包含 Layer、Compile 等)
        self.save_weights_only = save_weights_only
        if mode == 'min':
            # 设置 best 为无限大
            self.best = np.Inf
        else:
            # 设置 best 为负无限大
            self.best = -np.Inf

    # 存储网络模型的函数
    def save_model(self):
        if self.save_weights_only:
```

```python
            self.model.save_weights(self.weights_file)
        else:
            self.model.save(self.weights_file)

    def on_epoch_end(self, epoch, logs=None):
        # 从logs中读取监测值
        monitor_value = logs.get(self.monitor)
        # 如果监测值降低或升高（取决于mode设置），就存储网络模型
        if self.mode == 'min' and monitor_value < self.best:
            self.save_model()
            self.best = monitor_value
        elif self.mode == 'max' and monitor_value > self.best:
            self.save_model()
            self.best = monitor_value
```

- 建立回调函数

使用自定义回调函数存储模型权重。

```python
# 存储训练记录文件
logs_dirs = 'lab6-logs'
model_cbk = keras.callbacks.TensorBoard(log_dir='lab6-logs')

# 创建存储模型权重的目录
model_dirs = logs_dirs + '/models'
os.makedirs(model_dirs, exist_ok=True)
# 自定义回调函数存储模型权重
custom_save_model = SaveModel(model_dirs + '/custom_save.h5',
                              monitor='val_custom_catrgorical_accuracy',
                              mode='max',
                              save_weights_only=True)
```

- 建立 custom_categorical_crossentropy 损失函数

```python
def custom_categorical_crossentropy(y_true, y_pred):
    x = tf.nn.softmax_cross_entropy_with_logits(labels=y_true, logits=y_pred)
    return x
```

- 建立 CustomCategoricalAccuracy 评价指标函数

```python
class CustomCategoricalAccuracy(tf.keras.metrics.Metric):
    def __init__(self, name='custom_catrgorical_accuracy', **kwargs):
        super(CustomCategoricalAccuracy, self).__init__(name=name, **kwargs)
        # 记录正确预测的数量
        self.correct = self.add_weight('correct_numbers',
                                       initializer='zeros')
        # 记录全部数据的数量
        self.total = self.add_weight(total_numbers, initializer='zeros')

    def update_state(self, y_true, y_pred, sample_weight=None):
        # 输入答案为独热编码，所以取最大的数值为答案
        # EX:tf.argmax([[0,0,0,1,0,0,0,0,0,0]],axis=-1)=[4]
```

```
            y_true = tf.argmax(y_true, axis=-1)
            # 取预测输出最大的数值为预测结果
            y_pred = tf.argmax(y_pred, axis=-1)
            # 比较预测结果是否正确，正确会返回 True，错误会返回 False
            values = tf.equal(y_true, y_pred)
            # 转换为浮点数（True=1.0，False=0.0）
            values = tf.cast(values, tf.float32)
            values_sum = tf.reduce_sum(values)          # 将一个批量的数据加总
            num_values = tf.cast(tf.size(values), tf.float32)#计算这个批量的数据数量
            self.correct.assign_add(values_sum)         # 更新正确预测的总数
            self.total.assign_add(num_values)           # 更新数据量的总数

    def result(self):
        # 计算正确率
        return tf.math.divide_no_nan(self.correct, self.total)

    def reset_states(self):
        # 每一次 epoch 结束后会重新初始化变量
        self.correct.assign(0.)
        self.total.assign(0.)
```

- 设置训练使用的优化器、自定义损失函数和自定义评价指标函数

```
model_2.compile(keras.optimizers.Adam(),
                loss=custom_categorical_crossentropy,      # 自定义损失函数
                metrics=[CustomCategoricalAccuracy()])     # 自定义评价指标函数
```

- 训练网络模型

```
model_2.fit(train_data,
            epochs=100,
            validation_data=valid_data,
            callbacks=[model_cbk, custom_save_model])
```

结果如图 7-7 所示。

```
Epoch 96/100
704/704 [==============================] - 12s 17ms/step - loss: 0.7810 - custom_catrgorical_accuracy: 0.7374 - val_loss: 0.6471 - val_custom_catrgorical_accuracy: 0.7922
Epoch 97/100
704/704 [==============================] - 12s 17ms/step - loss: 0.7745 - custom_catrgorical_accuracy: 0.7422 - val_loss: 0.6365 - val_custom_catrgorical_accuracy: 0.7966
Epoch 98/100
704/704 [==============================] - 12s 17ms/step - loss: 0.7777 - custom_catrgorical_accuracy: 0.7420 - val_loss: 0.6332 - val_custom_catrgorical_accuracy: 0.7954
Epoch 99/100
704/704 [==============================] - 12s 17ms/step - loss: 0.7803 - custom_catrgorical_accuracy: 0.7416 - val_loss: 0.6337 - val_custom_catrgorical_accuracy: 0.8000
Epoch 100/100
704/704 [==============================] - 12s 17ms/step - loss: 0.7669 - custom_catrgorical_accuracy: 0.7473 - val_loss: 0.6622 - val_custom_catrgorical_accuracy: 0.7946
<tensorflow.python.keras.callbacks.History at 0x7f834551c208>
```

图 7-7　执行结果

Step 06 比较两种方法的训练结果。

（1）加载两种方法的模型权重：

```
model_1.load_weights(model_dirs+'/save.h5')
```

```
model_2.load_weights(model_dirs+'/custom_save.h5')
```

（2）验证网络模型：使用 Keras 高级 API 的正确率为 0.7897，而使用自定义 API 的正确率则为 0.8018。

```
loss_1, acc_1 = model_1.evaluate(test_data)
loss_2, acc_2 = model_2.evaluate(test_data)
loss = [loss_1, loss_2]
acc = [acc_1, acc_2]
dict = {"Loss": loss, "Accuracy": acc}
pd.DataFrame(dict)
```

结果如图 7-8 所示。

	Loss	Accuracy
0	0.661652	0.7897
1	0.627981	0.8018

图 7-8　执行结果

第 8 章

TensorBoard 高级技巧

学习目标

- 更深入地使用 TensorBoard，使用低级 API 记录数据，例如直方图、文字、图像等
- 在训练过程中，通过结合 tf.summary.image 和回调函数将图像记录到 TensorBoard
- 使用 TensorBoard 超参数调校工具来分析大量的训练模型

8.1 TensorBoard 的高级技巧

TensorBoard 是 TensorFlow 官方推出的可视化工具，在众多的机器学习工具中，TensorFlow 能成为如今很受欢迎的工具，TensorBoard 的功劳很大。在 3.4 节示范了 4 种可视化功能：SCALARS、GRAPHS、DISTRIBUTIONS 和 HISTOGRAMS 的使用，这 4 种功能只需要在训练时加入 tf.keras.callbacks.TensorBoard 回调函数即可完成设置。

但是 Keras 提供的 tf.keras.callbacks.TensorBoard 高级 API 能观察的重要指标有限，例如 SCALARS 中只有损失值（Loss）和指标值（Metrics）两种数值可以观察。如果想将输出结果用不同可视化工具进行分析并显示在 TensorBoard 上，就需要使用自定义函数来实现。

本章将说明 TensorBoard 的高级技巧，会先介绍如何通过低级的 tf.summary API 将想要记录的信息存储到 TensorBoard 中显示，内容如下：

- tf.summary：主要用来建立和记录 TensorBoard 记录文件。
- tf.summary.scaler：存储如损失、指标或学习率等的变化趋势。
- tf.summary.image：存储图像。
- tf.summary.text：存储一段文字。
- tf.summary.audio：存储可播放的音频。

- tf.summary.histogram：存储模型权重。

8.1.1　tf.summary

tf.summary 是 TensorFlow 提供的 TensorBoard 低级 API 指令，主要用来建立和把数据存储到 TensorBoard 记录文件中，最后可以利用 TensorBoard 可视化工具来读取记录文件中的信息。

- 建立记录文件

```
summary_writer = tf.summary.create_file_writer('lab7-logs-summary')
```

- 写入记录文件的方法

```
with summary_writer.as_default():  # summary_writer 是用于写入的默认记录文件
    tf.summary.(scalar | image | text | audio | histogram)(…)  # 写入的记录类型
```

- 启动 TensorBoard（命令行）查看训练记录（见图 8-1）。

```
tensorboard --logdir lab7-logs-summary
```

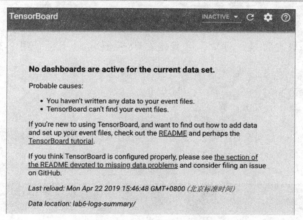

图 8-1　TensorBoard 页面

提　示
更多信息可以参阅 TensorFlow 官方 API 文件： https://www.tensorflow.org/versions/r2.0/api_docs/python/tf/summary。

8.1.2　tf.summary.scalar

tf.summary.scalar 为存储数值的函数，可以通过 TensorBoard SCALARS 可视化工具来显示存储的数值，如图 8-2 所示，显示网络模型训练过程的损失值和指标值。图 8-2 记录的损失值和指标值都是通过回调函数的 tf.keras.callbacks.TensorBoard 高级 API 来完成的。如果想要记录损失值和指标值之外的数值，就需要使用 tf.summary.scalar 低级 API 来完成。下面将以范例程序来说明 tf.summary.scalar API 如何使用。

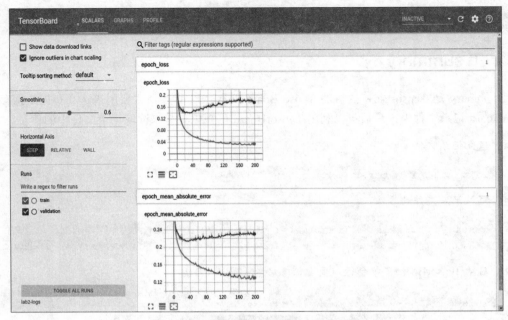

图 8-2　TensorBoard SCALARS

- 存储一个 sin 函数在记录文件中，并显示在 TensorBoard 上，如图 8-3 所示。

```
# 在 0~2π 之间线性产生 100 个点
x = np.linspace(0, 2 * np.pi , 100)
# 将 100 个点带入 sin 函数中计算出值
data = np.sin(x)
with summary_writer.as_default():  # summary_writer 是用于写入的默认记录文件
    for i, y in enumerate(data):
        tf.summary.scalar('sin', y, step=i)  # 存储，纵轴：数值 y，横轴：i 时间轴
```

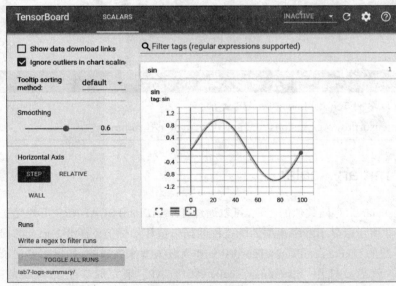

图 8-3　通过 TensorBoard SCALARS 观察 tf.summary.scalar API 的记录结果

8.1.3　tf.summary.image

tf.summary.image 为存储图像的函数，可以通过 TensorBoard IMAGES 可视化工具来显示存储的图像。下面将以范例程序来说明 tf.summary.image API 如何使用。

Step 01　到如下网址下载测试图像，并解压缩到当前的目录下：

　　　　https://drive.google.com/open?id=1cC45twI3a5AkBYYE6Qb3ZV3uCPlw9eL6

Step 02　从文件夹中读取一张图像（airplane.png）。

```
# 建立读取图像的函数
def read_img(file):
    image_string = tf.io.read_file(file)   # 读取文件
    # 将读入文件以图像格式来解码
    image_decode = tf.image.decode_image(image_string)
    return image_decode

img = read_img('image/airplane.png')    # 读入图像信息
plt.imshow(img)   # 显示读入的图像信息
```

结果如图 8-4 所示。

图 8-4　执行结果

Step 03　存储一张图像在记录文件中，并显示在 TensorBoard 上，如图 8-5 所示。

```
with summary_writer.as_default():  # summary_writer 是用于写入的默认记录文件
    tf.summary.image("Airplane", [img], step=0)   # 存入图像信息
```

Step 04　一次存储 5 张图像（airplane_zoom.png、airplane_flip.png、airplane_color.png、airplane_rot.png 和 airplane.png）到记录文件中，并显示在 TensorBoard 上，如图 8-6 所示。

```
img_files = [airplane_zoom.png, airplane_flip.png, airplane_color.png,
             airplane_rot.png, airplane.png]   # 建立一个用来存储读入图像的数组
imgs = []
for file in imgs:
    imgs.append(read_img('image/'+file))    # 读取图像并存入数组中

with summary_writer.as_default():  # summary_writer 是用于写入的默认记录文件
    # 一次存入 5 张图像（注意：如果 max_outputs 没有设置为 5，就只会存储 3 张图像）
```

```
tf.summary.image("Airplane Augmentation", imgs, max_outputs=5, step=0)
```

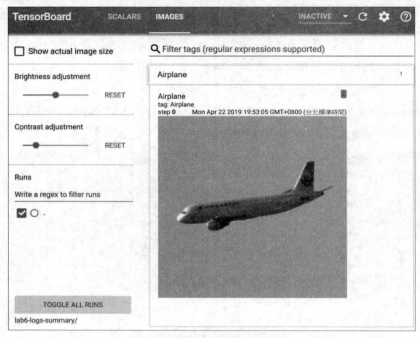

图 8-5　通过 TensorBoard IMAGES 观察 tf.summary.image API 记录的图像

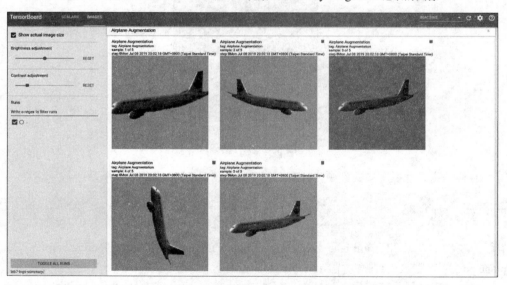

图 8-6　通过 TensorBoard IMAGES 观察 tf.summary.image API 记录的多张图像

Step 05　将 5 张图像（airplane_zoom.png、airplane_flip.png、airplane_color.png、airplane_rot.png 和 airplane.png）以不同 Step（时间）存储到记录文件中，并显示在 TensorBoard 上，如图 8-7 所示。

```
with summary_writer.as_default():  # summary_writer 是用于写入的默认记录文件
    # 每次存储 1 张图像，并存储在不同 Step 中
    for i, img in enumerate(imgs):
```

```
            tf.summary.image("Save image each step", [img], step=i)
```

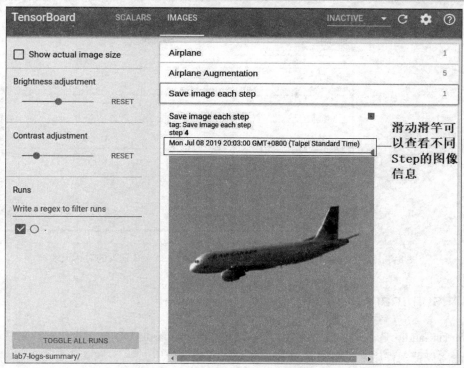

图 8-7　通过 TensorBoard IMAGES 观察 tf.summary.image API 以不同 Step 记录的多张图像

8.1.4　tf.summary.text

tf.summary.text 是存储文本数据的函数，可以通过 TensorBoard TEXT 可视化工具来显示存储的数据。下面将以范例程序来说明 tf.summary.text API 如何使用。

- 存储一段对话记录到记录文件中，并显示在 TensorBoard 上，如图 8-8 所示。

```
# 建立一个数组，里面包含对话记录
texts = ["小明：Cubee 小助理最近好想学深度学习的技术！",
         "Cubee：这是当然的了，这可是今天最火的技术呀！",
         "小明：那我该如何入门呢？",
         "Cubee：向你推荐一本书《轻松学会 TensorFlow 2.0 人工智能深度学习应用开发》。",
         "小明：这本书没有深度学习经验的人也能学会吗？",
         "Cubee：这是当然的，你只需要有基础的 Python 能力就可以学会了！",
         "小明：太好了，那我要赶快去买了！"]

with summary_writer.as_default():  # summary_writer 是用于写入的默认记录文件
    # 将每一段字符串信息以不同 Step 存入记录文件中
    for i, text in enumerate(texts):
        tf.summary.text("Chat record", text, step=i)
```

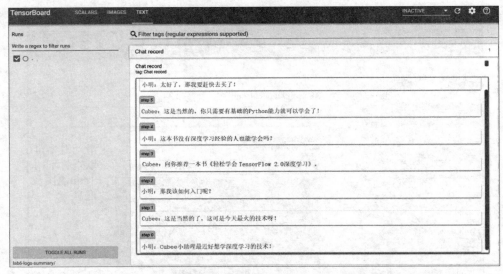

图 8-8　通过 TensorBoard TEXT 观察 tf.summary.text API 记录的对话记录

8.1.5　tf.summary.audio

tf.summary.audio 是存储音频的函数,可以通过 TensorBoard AUDIO 可视化工具来显示存储的音频。下面将以范例程序来说明 tf.summary.audio API 如何使用。

Step 01　到如下网址下载测试音频,并解压缩到当前目录下:

https://drive.google.com/open?id=1V4nNQ-ZQMBUezZEFWZoAZ62dORevUGPZ

Step 02　从文件夹(audio)中读取一个音频文件(cat.wav):

```
# 建立读取音频的函数
def read_audio(file):
    audio_string = tf.io.read_file(file)         # 读取文件
    # 将读入文件以音频格式来解码
    audio, fs = tf.audio.decode_wav(audio_string)
    # 因为tf.summary.audio要求输入格式为[k(clips), t(frames), c(channels)]
    # 编码后的音频只有[t(frames), c(channels)],所以音频需要增加一个维度
    audio = tf.expand_dims(audio, axis=0)
    return audio, fs

audio, fs = read_audio('./audio/cat.wav')    # 读取音频文件
```

Step 03　存储一个音频在记录文件中,并显示在 TensorBoard 上,如图 8-9 所示。

```
with summary_writer.as_default():    # summary_writer是用于写入的默认记录文件
    tf.summary.audio('cat', audio, fs, step=0)   # 存入音频信息
```

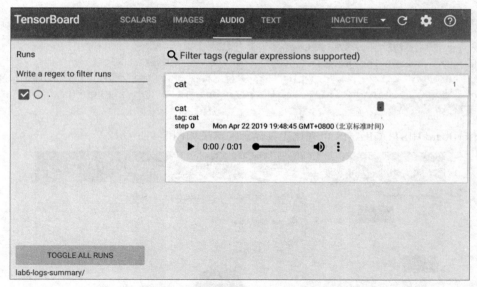

图 8-9　通过 TensorBoard AUDIO 观察 tf.summary.audio API 记录的音频

8.1.6　tf.summary.histogram

tf.summary.histogram 是存储数值分布的函数，可以通过 TensorBoard HISTOGRAMS 和 DISTRIBUTIONS 可视化工具来显示存储的数值分布。图 8-10 显示了网络模型各层的权重分布图，图中各网络层权重分布都是通过 tf.keras.callbacks.TensorBoard 高级 API 来完成的，但 tf.keras.callbacks.TensorBoard 仅提供网络层权重分布的记录功能。如果我们想要记录网络层的输出分布，就必须使用 tf.summary.histogram 低级 API 来完成。下面将以范例程序来说明 tf.summary.histogram API 如何使用。

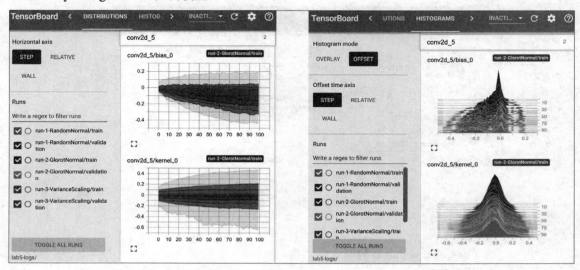

（a）TensorBoard DISTRIBUTIONS　　　　（b）TensorBoard HISTOGRAMS

图 8-10　网络模型各层的权重分布图

- 存储一个正态分布的数值在记录文件中，并显示在 TensorBoard 上

```
# 建立一个正态分布数据，数据共 64 笔，每笔有 100 个数据
data = tf.random.normal([64, 100], dtype=tf.float64)
with summary_writer.as_default():  # summary_writer 是用于写入的默认记录文件
    tf.summary.histogram('Normal distribution', data, step=0)
    # 存储正态分布的数据
```

TensorBoard HISTOGRAMS 可视化工具显示的结果如图 8-11 所示。

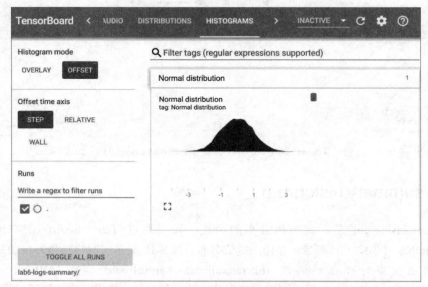

图 8-11　通过 TensorBoard HISTOGRAMS 观察 tf.summary.histogram API 记录的正态分布

TensorBoard DISTRIBUTIONS 可视化工具显示的结果如图 8-12 所示。

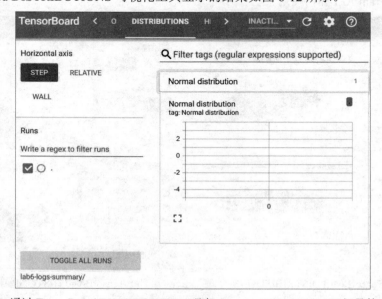

图 8-12　通过 TensorBoard DISTRIBUTIONS 观察 tf.summary.histogram API 记录的正态分布

第 8 章　TensorBoard 高级技巧

> **说　明**
> 因为 DISTRIBUTIONS 功能主要是用来看时间分布的变化，而上面的范例只存储了一个时间单位的数据，所以图 8-11 的 DISTRIBUTIONS 没有呈现任何结果。

- 存储 100 组正态分布的数据在记录文件，且每一组数据的平均值都相差 0.1

```
with summary_writer.as_default():   # summary_writer 是用于写入的默认记录文件
    # 存储 100 组正态分布的数据在记录文件中，且每一组数据的平均值都相差 0.1
    for i, offset in enumerate(tf.range(0, 10, delta=0.1, dtype=tf.float64)):
        tf.summary.histogram('Normal distribution 2', data+offset, step=i)
```

HISTOGRAMS 可视化工具显示的结果如图 8-13 所示。

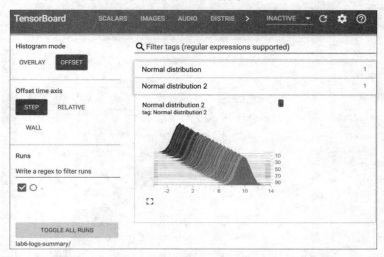

图 8-13　通过 TensorBoard HISTOGRAMS 观察 tf.summary.histogram API 记录多个正态分布变化的结果

DISTRIBUTIONS 可视化工具显示的结果如图 8-14 所示。

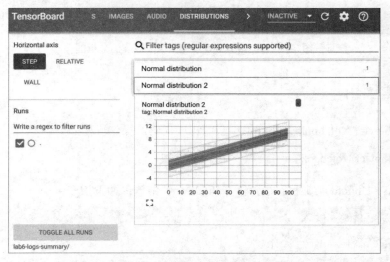

图 8-14　通过 TensorBoard DISTRIBUTIONS 观察 tf.summary.histogram API 记录多个正态分布变化的结果

8.2 实验一：使用 tf.summary.image 记录训练结果

为了让读者更熟悉 tf.summary 的功能，本实验会实现 CIFAR-10 范例，并使用混淆矩阵（Confusion Matrix，也称为误差矩阵）来分析网络模型，再利用 8.1 节学到的 tf.summary.image 将混淆矩阵存储起来，并在 TensorBoard 上使之可视化，如图 8-15 所示。

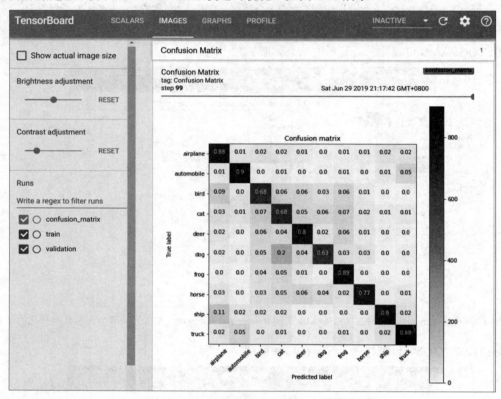

图 8-15　TensorBoard 可视化混淆矩阵的分析图

8.2.1 新建项目

建议使用 Jupyter Notebook 来执行本小节的程序代码，操作流程如下：

Step 01 启动 Jupyter Notebook。

在 Terminal（Ubuntu）或命令提示符（Windows）中输入如下指令：

```
jupyter notebook
```

Step 02 新建执行文件。

单击界面右上角的 New 下拉按钮，然后单击所安装的 Python 解释器（在 Jupyter 中都称为

Kernel）来启动它，如图 8-16 所示，显示了 3 个不同的 Kernel：

- Python3：本地端 Python。
- tf2：虚拟机 Python（TensorFlow-cpu 版本）。
- tf2-gpu：虚拟机 Python（TensorFlow-gpu 版本）。

图 8-16　新建执行文件

Step 03 执行程序代码。

按 Shift + Enter 快捷键执行单行程序代码，如图 8-17 所示。

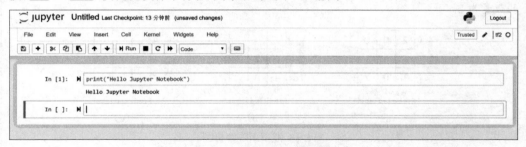

图 8-17　Jupyter 环境界面

接下来，后续的程序代码都可以在 Jupyter Notebook 上执行。

8.2.2　程序代码

Step 01 导入必要的套件。

```
import os
import io
import numpy as np
import tensorflow as tf
import tensorflow_datasets as tfds
import matplotlib.pyplot as plt
from tensorflow import keras
from tensorflow.keras import layers
# 从文件夹的 preprocessing.py 文件中导入（import）parse_aug_fn 和 parse_fn 函数
from preprocessing import parse_aug_fn, parse_fn
```

Step 02 计算混淆矩阵。

- 混淆矩阵函数

通过 tf.math.confusion_matrix 来计算混淆矩阵。

```
y_true = [2, 1, 0, 2, 2, 0, 1, 1]
y_pred = [0, 1, 0, 2, 2, 0, 2, 1]
cm = tf.math.confusion_matrix(y_true, y_pred, num_classes=3).numpy()
print(cm)
```

结果如下：

```
[[2 0 0]
 [0 2 1]
 [1 0 2]]
```

> **说明**
>
> 混淆矩阵的横轴代表真实标签，纵轴代表预测标签，如图 8-18（a）所示，图中数字代表的意思是预测为相对应类别的数量，例如图 8-18（b）中间的数字 2 代表真实类别为 1，并且正确预测 1 的数量为 2，图 8-18（c）左下角的数字 2 代表真实类别为 2，但错误预测成 0 的数量为 2。

图 8-18 混淆矩阵解析图

- 建立 plot_confusion_matrix 函数

将上面计算的混淆矩阵数组以 Matplotlib 图像来表示，而混淆矩阵中的数字改成百分比形式。

```
def plot_confusion_matrix(cm, class_names):
    """
    产生一张 Matplotlib 图像的混淆矩阵
    :param cm (shape = [n, n]): 传入混淆矩阵。
    :param class_names (shape = [n]): 传入类别名称。
    """
    # 归一化混淆矩阵
    cm = np.around(cm.astype('float') / cm.sum(axis=1)[:, np.newaxis], decimals=2)

    # 建立一个显示界面
    figure = plt.figure(figsize=(8, 8))
    # 根据 cm 的数值大小在界面中填入颜色
    plt.imshow(cm, interpolation='nearest', cmap=plt.cm.Blues)
    # 设置界面的标头文字
```

```python
plt.title("Confusion matrix")
# 刻度
tick_index = np.arange(len(class_names))
# Y 轴显示类别名称
plt.yticks(tick_index, class_names)
# X 轴显示类别名称,并将类别名称旋转 45 度(避免文字重叠)
plt.xticks(tick_index, class_names, rotation=45)
# 在图像右边产生一条颜色刻度条
plt.colorbar()

# 在每一格混淆矩阵输入预测百分比
threshold = cm.max() / 2.
for i in range(cm.shape[0]):
    for j in range(cm.shape[1]):
        # 如果格内背景颜色太深,就使用白色文字,反之使用黑色文字
        color = "white" if cm[i, j] > threshold else "black"
        plt.text(j, i, cm[i, j], horizontalalignment="center", color=color)
plt.ylabel('True label')
plt.xlabel('Predicted label')
# 对图像的位置进行调整,避免 x 或 y 轴的文字被遮挡
plt.tight_layout()
return figure

img = plot_confusion_matrix(cm, [0, 1, 2])
```

结果如图 8-19 所示。

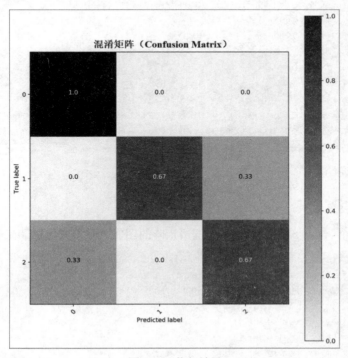

图 8-19　执行结果

> **说 明**
>
> **混淆矩阵归一化说明**
>
> 下面的程序为混淆矩阵归一化的过程，转换如图 8-20 所示。
>
> ```
> # np.around: decimals=2 代表取到小数点后第二位
> cm = np.around(cm.astype('float') / cm.sum(axis=1)[:, np.newaxis], decimals=2)
> ```
>
>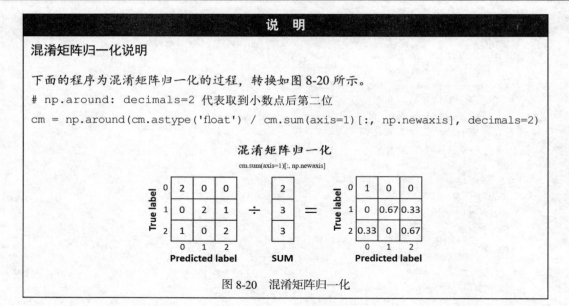
>
> 图 8-20 混淆矩阵归一化

- 建立 plot_to_image 函数

将 Matplotlib 图像转成 TensorFlow 形式的图像，这样才能够通过 tf.summary.image 把图像记录到 TensorBoard。

```python
def plot_to_image(figure):
    """将 Matplotlib 图像转成 TensorFlow 形式的图像"""
    # 将 Matplotlib plot 的图像以 PNG 的格式存储到内存中
    buf = io.BytesIO()
    plt.savefig(buf, format='png')
    # 关闭 plt 图像，防止图像直接显示在 Jupyter Notebook 界面中
    plt.close(figure)
    buf.seek(0)
    # 将内存中的数据转成 TensorFlow 形式的图像
    image = tf.image.decode_png(buf.getvalue(), channels=4)
    image = tf.expand_dims(image, 0)
    return image
```

Step 03 自定义回调函数。

- 建立回调函数

训练过程中每个 epoch 结束会产生一张混淆矩阵的图像，并将图像记录在 TensorBoard 上。

```python
class ConfusionMatrix(tf.keras.callbacks.Callback):
    def __init__(self, log_dir, test_data, class_name):
        super(ConfusionMatrix, self).__init__()
        self.log_dir = log_dir
        self.test_data = test_data
        self.class_names = class_name
        self.num_classes = len(class_name)

    def on_train_begin(self, self, logs=None):
```

```python
        path = os.path.join(self.log_dir, 'confusion_matrix')
        # 建立 TensorBoard 记录文件
        self.writer = tf.summary.create_file_writer(path)

    def on_epoch_end(self, epoch, logs=None):
        # 计算 Confusion matrix
        total_cm = np.zeros([10, 10])
        for x, y_true in self.test_data:
            y_pred = self.model.predict(x)
            y_pred = np.argmax(y_pred, axis=1)
            y_true = np.argmax(y_true, axis=1)
            cm = tf.math.confusion_matrix(y_true, y_pred,
                            num_classes=self.num_classes).numpy()
            total_cm += cm

        # 将混淆矩阵转成 Matplotlib 图像
        figure = plot_confusion_matrix(total_cm, class_names=
                                        self.class_names)
        # 将 Matplotlib 图像转成 TensorFlow 形式的图像
        cm_image = plot_to_image(figure)

        # 将图像记录在 TensorBoard log 中
        with self.writer.as_default():
            tf.summary.image("Confusion Matrix", cm_image, step=epoch)
```

Step 04 训练网络模型。

- 加载 CIFAR-10 数据集

```python
# 将训练数据重新分成 1:9 等分,分别分给验证数据和训练数据
valid_split, train_split = tfds.Split.TRAIN.subsplit([10, 90])
# 获取训练数据,以便读取 data 的信息
train_data, info = tfds.load("cifar10", split=train_split, with_info=True)
# 获取验证数据
valid_data = tfds.load("cifar10", split=valid_split)
# 获取测试数据
test_data = tfds.load("cifar10", split=tfds.Split.TEST)
# 获取 CIFAR-10 数据集的类别名称
class_name = info.features['label'].names
```

- 数据集设置

```python
AUTOTUNE = tf.data.experimental.AUTOTUNE    # 自动调整模式
batch_size = 64  # 批量大小
train_num = int(info.splits['train'].num_examples / 10) * 9  # 训练数据数量

train_data = train_data.shuffle(train_num)   # 打乱数据集
# 加载预处理 parse_aug_fn 函数,CPU 数量为自动调整模式
train_data = train_data.map(map_func=parse_aug_fn,
                            num_parallel_calls=AUTOTUNE)
# 设置批量大小,并启用 prefetch 模式(缓存空间为自动调整模式)
```

```
train_data = train_data.batch(batch_size).prefetch(buffer_size=AUTOTUNE)

# 加载预处理 parse_fn 函数,CPU 数量为自动调整模式
valid_data = valid_data.map(map_func=parse_fn, num_parallel_calls=AUTOTUNE)
# 设置批量大小,并启用 prefetch 模式(缓存空间为自动调整模式)
valid_data = valid_data.batch(batch_size).prefetch(buffer_size=AUTOTUNE)

# 加载预处理 parse_fn 函数,CPU 数量为自动调整模式
test_data = test_data.map(map_func=parse_fn, num_parallel_calls=AUTOTUNE)
# 设置批量大小,并启用 prefetch 模式(缓存空间为自动调整模式)
test_data = test_data.batch(batch_size).prefetch(buffer_size=AUTOTUNE)
```

- 建立网络模型

这里使用了以下几种网络层:

- keras.Input:输入层(输入图像的大小为 32×32×3)。
- layers.Conv2D:卷积层(使用 ReLU 激活函数,以及 3×3 大小的 Kernel),这里 Kernel 是指卷积核。
- layers.MaxPool2D:池化层(对特征图下采样)。
- layers.Flatten:压平层(特征图转成一维张量)。
- layers.Dropout:随机失活层(每次训练随机失活 50%的神经元)。
- layers.Dense:全连接层(隐藏层使用 ReLU 激活函数,最后一层使用线性输出)。

```
inputs = keras.Input(shape=(32, 32, 3))
x = layers.Conv2D(64, 3, activation='relu')(inputs)
x = layers.MaxPool2D()(x)
x = layers.Conv2D(128, 3, activation='relu')(x)
x = layers.Conv2D(256, 3, activation='relu')(x)
x = layers.Conv2D(128, 3, activation='relu')(x)
x = layers.Conv2D(64, 3, activation='relu')(x)
x = layers.Flatten()(x)
x = layers.Dense(64, activation='relu')(x)
x = layers.Dropout(0.5)(x)
outputs = layers.Dense(10)(x)
# 建立网络模型(将输入到输出所有经过的网络层连接起来)
model_1 = keras.Model(inputs, outputs, name='model_1')
model_1.summary()
```

结果如图 8-21 所示。

第 8 章 TensorBoard 高级技巧

```
Model: "model_1"
_____
Layer (type)                 Output Shape              Param #
=================================================================
input_1 (InputLayer)         [(None, 32, 32, 3)]       0
_____
conv2d (Conv2D)              (None, 30, 30, 64)        1792
_____
max_pooling2d (MaxPooling2D) (None, 15, 15, 64)        0
_____
conv2d_1 (Conv2D)            (None, 13, 13, 128)       73856
_____
conv2d_2 (Conv2D)            (None, 11, 11, 256)       295168
_____
conv2d_3 (Conv2D)            (None, 9, 9, 128)         295040
_____
conv2d_4 (Conv2D)            (None, 7, 7, 64)          73792
_____
flatten (Flatten)            (None, 3136)              0
_____
dense (Dense)                (None, 64)                200768
_____
dropout (Dropout)            (None, 64)                0
_____
dense_1 (Dense)              (None, 10)                650
=================================================================
Total params: 941,066
Trainable params: 941,066
Non-trainable params: 0
```

图 8-21　执行结果

- 建立回调函数

```
# 存储训练记录文件
logs_dirs = 'lab7-logs-images'
model_cbk = keras.callbacks.TensorBoard(logs_dirs)
# 存储混淆矩阵图像
save_cm = ConfusionMatrix(logs_dirs, test_data, class_name)
```

- 设置训练使用的优化器、损失函数和评价指标函数

```
model_1.compile(keras.optimizers.Adam(),
        loss=keras.losses.CategoricalCrossentropy(from_logits=True),
        metrics=[keras.metrics.CategoricalAccuracy()])
```

- 训练网络模型

```
model_1.fit(train_data,
        epochs=100,
        validation_data=valid_data,
        callbacks=[model_cbk, save_cm])
```

结果如图 8-22 所示。

```
Epoch 96/100
704/704 [==============================] - 13s 18ms/step - loss: 0.7863 - categorical_accuracy: 0.7395 - val_loss: 0.6702 - val_categorical_accuracy: 0.7900
Epoch 97/100
704/704 [==============================] - 13s 18ms/step - loss: 0.7890 - categorical_accuracy: 0.7370 - val_loss: 0.6618 - val_categorical_accuracy: 0.7960
Epoch 98/100
704/704 [==============================] - 13s 19ms/step - loss: 0.7761 - categorical_accuracy: 0.7423 - val_loss: 0.6465 - val_categorical_accuracy: 0.7918
Epoch 99/100
704/704 [==============================] - 13s 18ms/step - loss: 0.7992 - categorical_accuracy: 0.7354 - val_loss: 0.6544 - val_categorical_accuracy: 0.7866
Epoch 100/100
704/704 [==============================] - 13s 18ms/step - loss: 0.7820 - categorical_accuracy: 0.7400 - val_loss: 0.6489 - val_categorical_accuracy: 0.7966
<tensorflow.python.keras.callbacks.History at 0x7f28b0183da0>
```

图 8-22　执行结果

Step 05 TensorBoard 观察存储的混淆矩阵。

- 启动 TensorBoard（命令行）查看训练记录

```
tensorboard --logdir lab7-logs-images
```

- 使用 TensorBoard 观察混淆矩阵

通过 TensorBoard 观察训练中混淆矩阵的变化，可以拖曳进度条来观察不同 epoch 的网络预测结果，如图 8-23 所示。

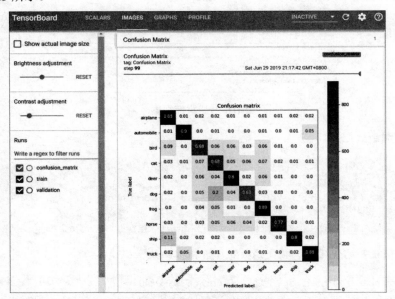

图 8-23　通过 TensorBoard 观察混淆矩阵 1

如图 8-24 所示，小方框内的数值越大，就代表网络的性能越好。

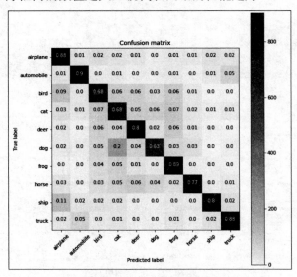

图 8-24　通过 TensorBoard 观察混淆矩阵 2

另外，从混淆矩阵中还可以观察出类别之间的关系，如图 8-25 所示，输入猫的图像进去网络预测，会有 20%的概率预测为狗，这意味着网络很容易将猫误认为狗。

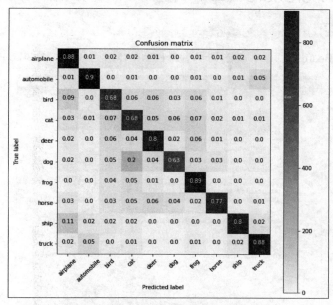

图 8-25　通过 TensorBoard 观察混淆矩阵 3

8.3　实验二：使用 TensorBoard 超参数调校工具来训练多个网络模型

本节来介绍更高级的工具——TensorBoard 超参数调校工具。在建立训练模型时，开发者常会面临要建立多少层的网络、使用哪一种优化器、学习率要调整多少、要不要加入 Dropout（随机失活）或批量归一化等问题，而这些可调整的参数统称为超参数。超参数的组合非常多，尽管可以筛选出一些必要的超参数进行测试并分析哪些超参数对训练模型有帮助，但是这么做效益并不高。因此，本节将介绍 TensorBoard 提供的超参数调校工具来解决上述问题，下面将直接以范例程序来说明。

本节的范例程序在 CIFAR-10 数据集上训练了 36 组不同超参数的模型进行分析比较，这 36（2×2×3×3）种组合分别如下：

- 图像增强：是或否。
- 批量归一化：是或否。
- 学习率：0.001、0.01、0.03。
- 初始化方法：Random Normal、Glorot Normal、He Normal。

8.3.1 启动 TensorBoard（命令行）

```
tensorboard --logdir lab7-logs-hparams
```

Step 01 打开 TensorBoard 网页后，单击列表中的 HPARAMS 选项，如图 8-26 所示。

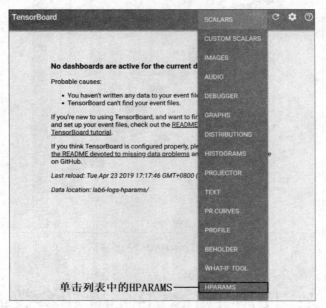

图 8-26　TensorBoard 的列表选项

Step 02 转至 Hyperparameters 可视化工具页面，如图 8-27 所示，因为还未建立记录文件，所以这个页面并没有任何信息显示。

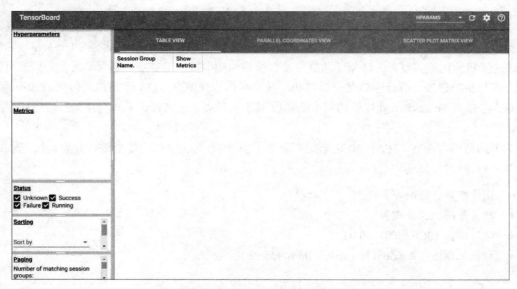

图 8-27　Hyperparameters 页面

8.3.2 程序代码

Step 01 导入必要的套件。

```
Import os
import tensorflow as tf
import tensorflow_datasets as tfds
from tensorflow import keras
from tensorflow.keras import layers
from tensorflow.keras import initializers
# 从文件夹的 preprocessing.py 文件中导入 parse_aug_fn 和 parse_fn 函数
from preprocessing import parse_aug_fn, parse_fn
```

Step 02 导入 TensorBoard 超参数工具所需要的套件。

```
from tensorboard.plugins.hparams import api as hp
```

Step 03 设置 TensorBoard 超参数调校。

定义要测试的超参数，总共 36（2×2×3×3）种组合：

- 图像增强：是或否。
- 批量归一化：是或否。
- 学习率：0.001、0.01、0.03。
- 初始化方法：Random Normal、Glorot Normal、He Normal。

```
hparam_ia = hp.HParam('Imgae_Augmentation', hp.Discrete([False, True]))
hparam_bn = hp.HParam('Batch_Normalization', hp.Discrete([False, True]))
hparam_init = hp.HParam('Weight_Initialization',
                        hp.Discrete(['RandomNormal_0.01std',
                                     'glorot_normal',
                                     'he_normal']))
hparam_lr = hp.HParam('Learning_Rate', hp.Discrete([0.001, 0.01, 0.03]))
```

Step 04 将实验摘要写入记录文件。

设置实验超参数信息和指标信息：

```
# 建立 TensorBoard 日志文件
logs_dirs = os.path.join('lab7-logs-hparams', 'hparam_tuning')
root_logdir_writer = tf.summary.create_file_writer(logs_dirs)
with root_logdir_writer.as_default():
# root_logdir_writer 是用于写入的默认记录文件
# 将实验超参数信息和指标信息写到 TensorBoard 记录文件中
hp.hparams_config(hparams=[hp_ia, hp_bn, hp_init, hp_lr], metrics
= [hp_metric])

# 实验指标信息
metric = 'Accuracy'
# 建立 TensorBoard 日志文件
```

```
log_dirs = "lab7-logs-hparams/hparam_tuning"
with tf.summary.create_file_writer(log_dirs).as_default():
    # 将实验超参数信息和指标信息写到 TensorBoard 记录文件中
    hp.hparams_config(hparams=[hparam_ia, hparam_bn, hparam_init, hparam_lr],
metrics=[hp.Metric(metric, display_name='Accuracy')],
)
```

执行完上面的程序后，我们会发现 TensorBoard 显示界面多了超参数和指标信息，如图 8-28 所示。

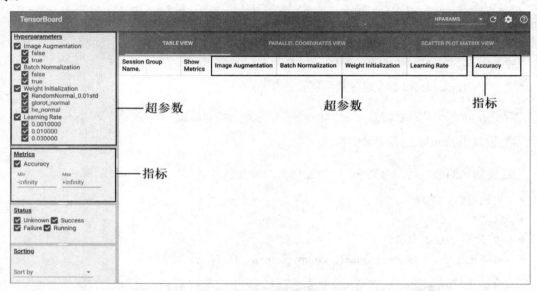

图 8-28　TensorBoard HPARAMS 超参数和指标信息

Step 05 训练网络模型。

加载 CIFAR-10 数据集：

```
# 获取训练数据
valid_split, train_split = tfds.Split.TRAIN.subsplit([10, 90])
# 获取训练数据(没有经过图像增强)
train_data_noaug, info = tfds.load("cifar10", split=train_split,
                                   with_info=True)
# 获取训练数据 (经过图像增强)
train_data_aug = tfds.load("cifar10", split=train_split)
# 获取验证数据
valid_data = tfds.load("cifar10", split=valid_split)
```

准备训练数据，分别为没有经过图像增强和经过图像增强的训练数据：

```
# 对没有经过图像增强的训练数据进行设置
train_data_noaug = train_data_noaug.shuffle(train_num)  # 打乱数据集
# 加载预处理 parse_fn 函数，CPU 数量为自动调整模式
train_data_noaug=train_data_noaug.map(map_func=parse_fn,
                                      num_parallel_calls=AUTOTUNE)
# 设置批量大小，并启用 prefetch 模式（缓存空间为自动调整模式）
```

```python
train_data_noaug=train_data_noaug.batch(batch_size)
                              .prefetch(buffer_size=AUTOTUNE)

# 对经过图像增强的训练数据进行设置
train_data_aug = train_data_aug.shuffle(train_num)  # 打乱数据集
# 加载预处理 parse_aug_fn 函数，CPU 数量为自动调整模式
train_data_aug=train_data_aug.map(map_func=parse_aug_fn,
                              num_parallel_calls=AUTOTUNE)
# 设置批量大小，并启用 prefetch 模式（缓存空间为自动调整模式）
train_data_aug = train_data_aug.batch(batch_size)
                              .prefetch(buffer_size=AUTOTUNE)

# 加载预处理 parse_fn 函数，CPU 数量为自动调整模式
valid_data = valid_data.map(map_func=parse_fn, num_parallel_calls=AUTOTUNE)
# 设置批量大小，并启用 prefetch 模式（缓存空间为自动调整模式）
valid_data = valid_data.batch(batch_size).prefetch(buffer_size=AUTOTUNE)
```

自定义超参数回调函数，用于记录每一次训练模型使用的超参数和最终训练结果的指标值：

```python
class HyperparameterCallback(tf.keras.callbacks.Callback):
    # 类别建立时调用
    def __init__(self, log_dir, hparams):
        super(HyperparameterCallback, self).__init__()
        self.log_dir = log_dir
        self.hparams = hparams
        self.best_accuracy = 0
        self.writer = None

    # 训练开始前调用，建立一个记录文件
    def on_train_begin(self, logs=None):
        self.writer = tf.summary.create_file_writer(self.log_dir)

    # 每一个 epoch 结束后调用，每一个训练 epoch 结束后，如果模型有进步，就会更新正确率
    def on_epoch_end(self, epoch, logs=None):
        current_accuracy = logs.get('val_categorical_accuracy')
        if current_accuracy > self.best_accuracy:
            self.best_accuracy = current_accuracy

    # 训练结束时调用，将训练超参数和最佳的正确率存入记录文件中
    def on_train_end(self, logs=None):
        with self.writer.as_default():
            hp.hparams(self.hparams)  # 记录这一次训练的权重参数
            tf.summary.scalar('accuracy', self.best_accuracy, step=0)
```

建立一个函数负责建立、编译和训练网络模型，并且会依照传入的超参数调整网络。网络层配置如下：

- keras.Input：输入层（输入图像的大小为 32×32×3）。
- layers.Conv2D：卷积层使用 3×3 大小的 Kernel（超参数：正态分布（std 0.01）初始化、Glorot 初始化或 He 初始化），这里 Kernel 是指卷积核。

- layers.BatchNormalization：BatchNormalization 层使用默认参数（超参数：使用或不使用）。
- layers.ReLU：ReLU 激活函数层。
- layers.MaxPool2D：池化层（对特征图下采样）。
- layers.Flatten：压平层（特征图转成一维张量）。
- layers.Dropout：随机失活层（每次训练随机失活 50%的神经元）。
- layers.Dense：全连接层（隐藏层使用 ReLU 激活函数，输出层使用 Softmax 激活函数）。

```python
def train_test_model(logs_dir, hparams):
    """
    logs_dir:传入目前执行的任务日志文件的位置
    hparams:传入超参数
    """
    # 权重初始化：初始化使用 Glorot、He 或标准差为 0.01 的正态分布
    if hparams[hp_init] == "glorot_normal":
        init = initializers.glorot_normal()
    elif hparams[hp_init] == "he_normal":
        init = initializers.he_normal()
    else:
        init = initializers.RandomNormal(0, 0.01)

    inputs = keras.Input(shape=(32, 32, 3))
    x = layers.Conv2D(64, (3, 3), kernel_initializer=init)(inputs)
    # 批量归一化层：选择加或不加批量归一化层
    if hparams[hp_bn]: x = layers.BatchNormalization()(x)
    x = layers.ReLU()(x)
    x = layers.MaxPool2D()(x)
    x = layers.Conv2D(128, (3, 3), kernel_initializer=init)(x)
    # 批量归一化层：选择加或不加批量归一化层
    if hparams[hp_bn]: x = layers.BatchNormalization()(x)
    x = layers.ReLU()(x)
    x = layers.Conv2D(256, (3, 3), kernel_initializer=init)(x)
    # 批量归一化层：选择加或不加批量归一化层
    if hparams[hp_bn]: x = layers.BatchNormalization()(x)
    x = layers.ReLU()(x)
    x = layers.Conv2D(128, (3, 3), kernel_initializer=init)(x)
    # 批量归一化层：选择加或不加批量归一化层
    if hparams[hp_bn]: x = layers.BatchNormalization()(x)
    x = layers.ReLU()(x)
    x = layers.Conv2D(64, (3, 3), kernel_initializer=init)(x)
    # 批量归一化层：选择加或不加批量归一化层
    if hparams[hp_bn]: x = layers.BatchNormalization()(x)
    x = layers.ReLU()(x)
    x = layers.Flatten()(x)
    x = layers.Dense(64, kernel_initializer=init)(x)
    # 批量归一化层：选择加或不加批量归一化层
    if hparams[hp_bn]: x = layers.BatchNormalization()(x)
    x = layers.ReLU()(x)
    x = layers.Dropout(0.5)(x)
    outputs = layers.Dense(10, activation='softmax')(x)
```

```python
# 建立网络模型（将输入到输出所有经过的网络层连接起来）
model = keras.Model(inputs, outputs, name='model')

# 存储训练记录文件
model_tb = keras.callbacks.TensorBoard(log_dir=logs_dir,
                                        write_graph=False)

# 存储最好的网络模型权重
model_mckp = keras.callbacks.ModelCheckpoint(logs_dir +'/best-model.h5',
                            monitor='val_categorical_accuracy',
                            save_best_only=True,
                            mode='max')

# 设置停止训练的条件（如果正确率超过 30 次迭代都没有上升，就终止训练）
model_els = keras.callbacks.EarlyStopping
                    (monitor='val_categorical_accuracy',
                    min_delta=0,
                    patience=30,
                    mode='max')
# 自定义超参数回调函数，记录训练模型的超参数和指标（正确率）
model_hparam = HyperparameterCallback(logs_dir + 'validation', hparams)

# 设置训练使用的优化器、损失函数和评价指标函数
# 优化器学习率为超参数：0.001、0.01 或 0.03
model.compile(keras.optimizers.Adam(hparams[hp_lr]),
            loss=keras.losses.CategoricalCrossentropy(),
            metrics=[keras.metrics.CategoricalAccuracy()])

# 超参数：使用经过图像增强的数据或没有经过图像增强的数据训练网络
if hparams[hp_ia]:
    history = model.fit(train_data_aug,
                epochs=100,
                validation_data=valid_data,
                callbacks=[model_tb,model_mckp,model_els,model_hparam])
else:
    history = model.fit(train_data_noaug,
                epochs=100,
                validation_data=valid_data,
                callbacks=[model_tb,model_mckp,model_els,model_hparam])
```

训练 36 种不同参数的网络模型。

```python
session_id = 1   # 训练任务的 id
# 存储记录文件的位置
logs_dir = os.path.join('lab7-logs-hparams', 'run-{}')
for ia in ia_list:
    for bn in bn_list:
        for init in init_list:
            for lr in lr_list:
                # 显示目前训练任务 id
                print('--- Running training session {}'.format(session_id))
```

```
                # 设置本次训练的超参数
                hparams = {hp_ia: ia, hp_bn: bn, hp_init: init, hp_lr: lr}
                # 建立、编译及训练网络模型
                train_test_model(logs_dir.format(session_id), hparams)
                session_id += 1  # id+1
```

Step 06 通过 TensorBoard 超参数调校工具分析训练模型。

完成训练后,TensorBoard 会显示所有训练任务的超参数和指标,如图 8-29 所示。

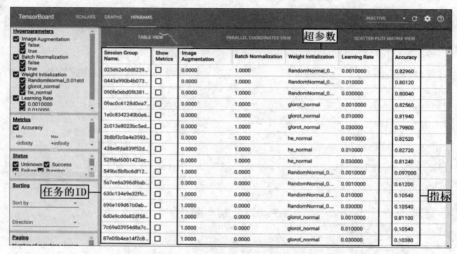

图 8-29　TensorBoard HPARAMS 记录了 36 组训练的超参数和指标

- **PARALLEL COORDINATES VIEW 功能**

(1) 单击 PARALLEL COORDINATES VIEW 按钮,如图 8-30 所示。

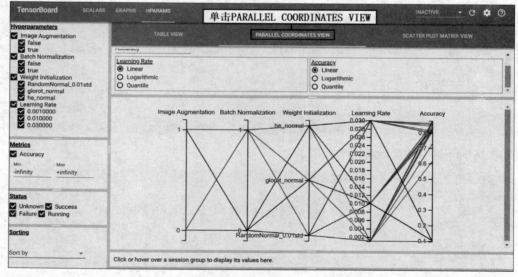

图 8-30　TensorBoard HPARAMS PARALLEL COORDINATES VIEW 页面

（2）框出感兴趣的正确率范围，显示界面会自动将正确率范围内的模型标示出来，如图 8-31 所示。我们可以从图中发现两个特性：高正确率的模型都有经过图像增强和批量归一化；使用哪一种初始化或学习率对训练结果影响不大。

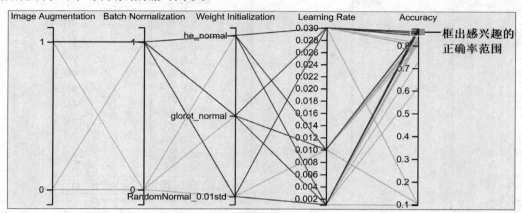

图 8-31　TensorBoard HPARAMS PARALLEL COORDINATES VIEW 操作

（3）可以加入指标的范围限制，这样可以过滤不必要的信息，如图 8-32 所示。从图中可以观察出，如果有批量归一化，权重初始化的重要性就不大了。

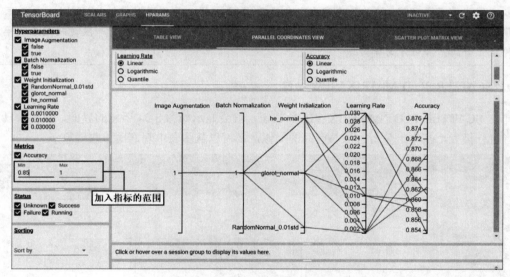

图 8-32　TensorBoard HPARAMS 指标的范围限制

（4）将 Learning Rate 0.001 的设置框起来，可以更清楚地发现最高正确率的 3 组模型都是使用 0.001 学习率进行训练的，如图 8-33 所示。

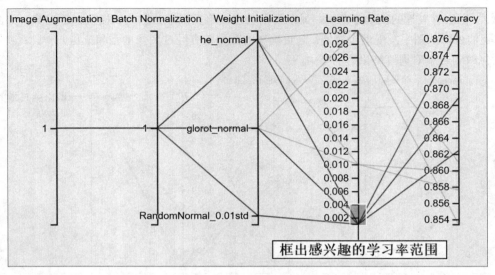

图 8-33　TensorBoard HPARAMS PARALLEL COORDINATES VIEW 学习率操作

说　明
图 8-33 的测试结果显示学习率 0.001 会训练出最好的模型，但这并不是一个非常公平的比较，因为学习率 0.001 的收敛速度远比 0.01 和 0.03 要慢，且使用 ReduceLROnPlateau 等回调函数在训练过程中会降低学习率，所以可以尝试提高学习率，每经过一段时间或正确率停止上升时，降低学习率。

- SCATTER PLOT MATRIX VIEW 功能

单击 SCATTER PLOT MATRIX VIEW 按钮，就会显示如图 8-34 所示的界面，这部分功能留给读者自行摸索，基本上分析方法与前面的类似，都可以从图表中观察出一些现象。

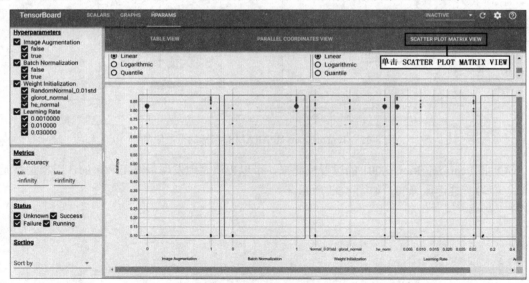

图 8-34　TensorBoard HPARAMS SCATTER PLOT MATRIX VIEW 页面

# 第 9 章

卷积神经网络经典架构

学习目标

- 认识 LeNet、AlexNet、VGG、GoogLeNet 和 ResNet 等代表性的网络架构
- 通过 Keras Applications 来实现 Inception V3 网络架构
- 通过 TensorFlow Hub 来实现 Inception V3 网络架构

9.1 神经网络架构

本节要介绍卷积神经网络历史中具有代表性的几个网络架构,依序分别为 LeNet（1998）[1]、AlexNet（2012）[2]、VGG（2014）[3]、GoogLeNet（2014）[4]和 ResNet（2015）[9]，以上架构都可以应用于人脸识别、图像分类等。

9.1.1 LeNet

LeNet[1]由 Yann Le Cun 于 1998 年提出，应用于手写数字识别，是卷积神经网络的始祖，而 LeNet 与现在的卷积神经网络最大的差别在于，隐藏层的激活函数使用 Sigmoid 而非 ReLU。网络架构模型如图 9-1 所示，包含两层卷积层和三层全连接层。

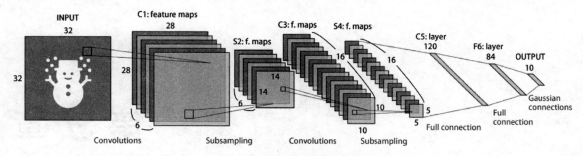

图 9-1　LeNet 网络模型

9.1.2　AlexNet

AlexNet[2]是 2012 年 ImageNet 大赛的冠军，网络模型如图 9-2 所示，它的网络架构比 LeNet 还要大很多，且隐藏层激活函数改成了 ReLU，这让网络可以训练更深的架构，此外还加入了 Dropout 层。如果说 LeNet 是卷积神经网络的始祖，那 AlexNet 就是带动深度学习热潮的关键。

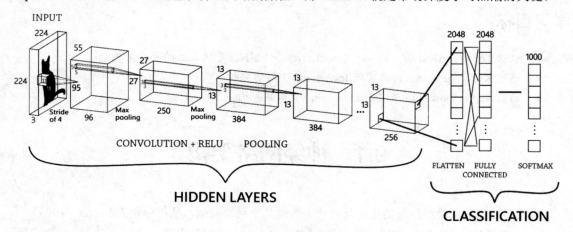

图 9-2　AlexNet 网络模型

9.1.3　VGG

VGG[3]是由 Visual Geometry Group 牛津大学科学工程系提出的网络架构，它有非常多的版本，例如 VGG11、VGG13、VGG16 和 VGG19，各版本的差异为卷积层的数量，网络架构主要是卷积层加全连接层，图 9-3 所示为 VGG16 网络模型。

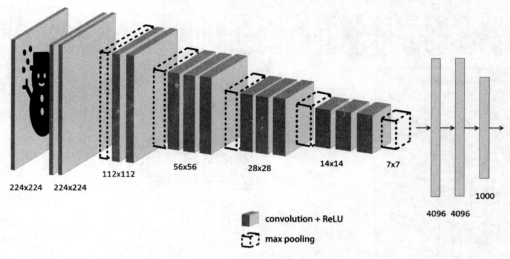

图 9-3　VGG16 网络模型

　　VGG 的特点是卷积层的卷积核都使用 3×3 的大小，主要原因是使用多个 3×3 的卷积层堆叠一样可以达到 5×5、7×7、9×9 相同的感受野（Receptive Field），但使用的参数量却可以大幅减少。例如，图 9-4（a）两个 3×3 卷积层和图 9-4（b）单个 5×5 卷积层是相同的输入大小和输出大小，但使用两个 3×3 卷积层只需要 18 个参数，而单个 5×5 卷积层则需要 25 个参数。

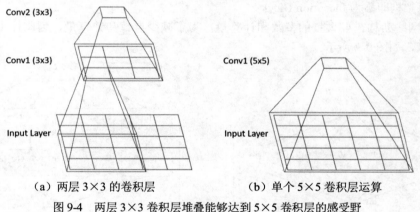

（a）两层 3×3 的卷积层　　　　　　（b）单个 5×5 卷积层运算

图 9-4　两层 3×3 卷积层堆叠能够达到 5×5 卷积层的感受野

9.1.4　GoogLeNet

　　GoogLeNet[4]是由 Google 提出的网络架构，又称作 Inception V1，是 2014 年 ImageNet 的冠军，为了向 LeNet 致敬，因此 GoogLeNet 名字是由 Google 加上 LeNet 缩写而成的，网络模型如图 9-5 所示。

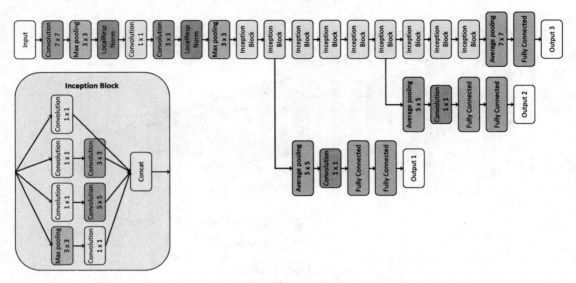

图 9-5　GoogLeNet 网络模型

GoogLeNet 最大的贡献在于提出了全新的架构 Inception Block，最初设计 Inception 的初衷是为了免去搭建网络架构时需要思考使用"几×几"的卷积核，或者是否使用池化层的烦恼，故它的原版设计非常简单且直观，对每一层的输出都做了 1×1、3×3、5×5 的卷积和 3×3 的池化。如图 9-6 所示，该架构即称为 Inception Block。

上面的网络架构需要大量的参数和计算量，为了减少参数和计算量，因此提出了 Inception Block 改进版，如图 9-7 所示。

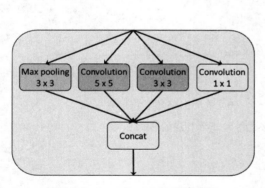

图 9-6　Inception Block 简易版

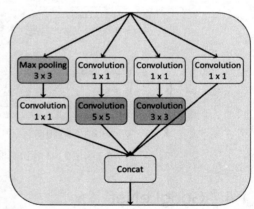

图 9-7　Inception Block 改进版

图 9-6 与图 9-7 的差别在于，3×3 和 5×5 的卷积层前多了一个 1×1 的卷积层，以及在 3×3 的池化后面也多了一个 1×1 的卷积层。由于加入了 1×1 的卷积层这个设计，因此可以大量减少参数与计算量。例如，输入大小为 36×36×128，希望经过卷积后得到 36×36×64 大小的输出。图 9-9 使用 Inception Block 改进版的参数和计算量比图 9-8 使用 Inception Block 简易版要少好几倍。

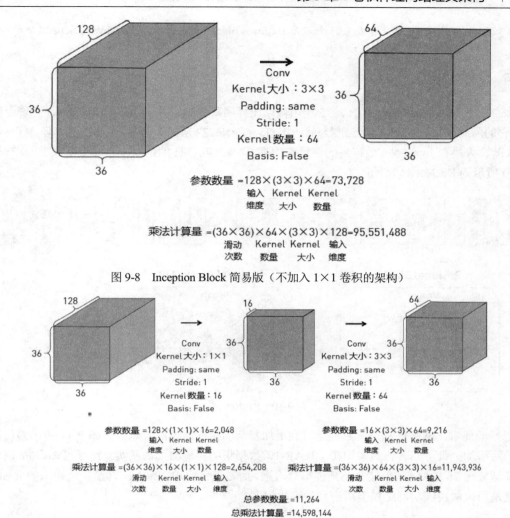

图 9-8　Inception Block 简易版（不加入 1×1 卷积的架构）

图 8-9　Inception Block 改进版（加入 1×1 卷积的架构）

GoogLeNet 由 9 个 Inception Block 组成，网络深度虽然有 22 层，但参数比 AlexNet 8 层少了 12 倍。在深度学习发展历程中，GoogLeNet 是一个很重要的里程碑，与前面提到的网络架构相比，GoogLeNet 在网络架构上有更多变化。

Google 研究团队在提出 GoogLeNet 架构之后，还推出了许多 Inception 的版本，各版本的简介如下：

- GoogLeNet（又称作 Inception-V1）[4]：于 2014 年发表，为首个多支路架构，引入了 1×1 的卷积层减少网络计算量。
- Inception-V2[5]：于 2015 年发表，引入了批量归一化，并使用两个 3×3 的卷积替代 5×5 的卷积。
- Inception-V3[6]：于 2016 年发表，使用 1×n 和 n×1 的卷积替代 n×n 的卷积。
- Inception-V4[7]：于 2017 年发表，引入了 ResNet 的 Shortcut（捷径）概念。

- Xception[8]：于 2017 年发表，全名为 Extreme Inception，引入了 Depthwise Separable Conv。

9.1.5 ResNet

ResNet[9]（Residual Nets）于 2015 年由微软研究团队提出，并在当年的 ImageNet 竞赛上获得冠军，它的网络层数共为 152 层，深度为前一年 GoogLeNet 22 层的 7 倍多，并且错误率降到了 3.6%。根据实验，人类在 ImageNet 数据集的错误率约为 5.1%，ResNet 的错误率已经小于人类的错误率。图 9-10 所示为 ResNet 34 层的版本。

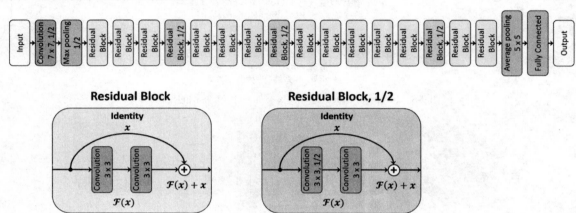

图 9-10　ResNet-34

以经验而言，网络层越深，越能学习到更加复杂的特征，性能就越好。图 9-11 所示为使用卷积神经网络加上批量归一化的架构在 CIFAR-10 数据集的实验结果，从实验结果可知，56 层网络的训练结果比 26 层要差，但这并不是过拟合问题，而是退化问题。ResNet 这篇论文提出的 Residual 架构就是用来解决深层网络模型所遇到的退化问题。

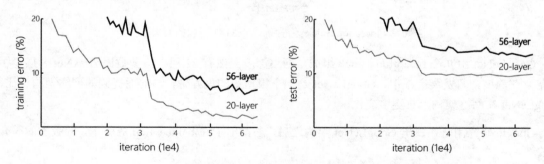

图 9-11　20 层网络与 56 层网络在 CIFAR-10 数据集上的误差

一个典型的 Residual 架构（或称作 Residual Block）如图 9-12 所示，架构关键在于 Identity（或称为 Shortcut，捷径）。至于为什么加上这种机制会对深层网络有帮助，这里举一个较为容易理解的例子：小时候玩过传话游戏，人数越多，游戏难度就越高，其中人数就好比网络的层数，当网络层数在 8 层的时候，已经能够得到最佳解，后面多出来的网络层反而会影响最佳解的传播，所以 Identity 的功能就是直接将第 8 层的输出传到最后。

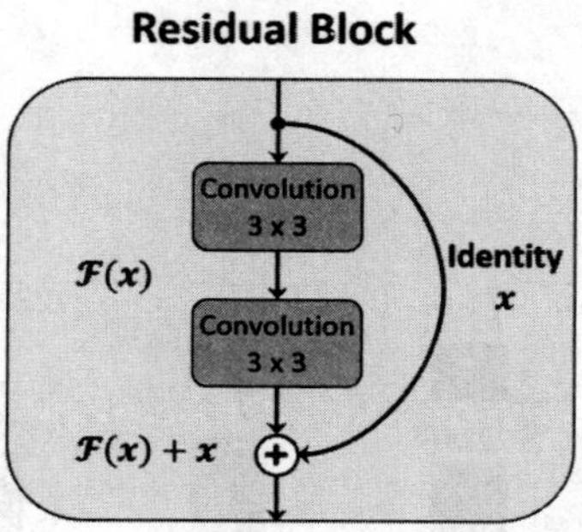

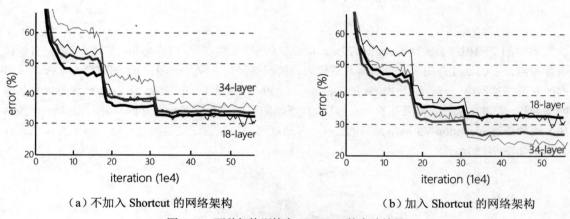

图 9-12 Residual Block

图 9-13 所示为论文中使用 ResNet 在 ImageNet 数据集上的实验结果，图 9-13（a）的训练曲线图为不加入 Shortcut 机制，可观察出 34 层网络的训练结果比 18 层还要差；图 9-13（b）的训练曲线图为加入 Shortcut 机制，也就是使用典型的 Residual Block 架构，可观察出 34 层的训练结果优于 18 层。

（a）不加入 Shortcut 的网络架构　　　　　　（b）加入 Shortcut 的网络架构

图 9-13 两种架构训练在 ImageNet 的实验结果

9.1.6　总结各种网络架构的比较

图 9-14 所示为前面介绍的 AlexNet、VGG、GoogLeNet 和 ResNet 四种网络架构在 ImageNet Large Scale Visual Recognition Challenge（ILSVRC）图像分类挑战赛上的 Top5 错误率和网络层的深度对比图。

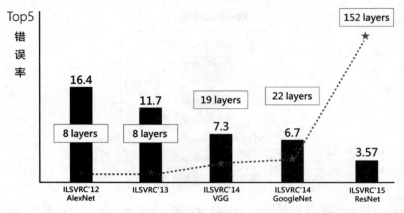

图 9-14　AlexNet、VGG、GoogLeNet 和 ResNet 的 Top5 错误率和网络层的深度对比图

9.2　实验：实现 Inception V3 网络架构

本节的范例程序会实现 Google 的 Inception V3 网络架构，并用 Inception V3 网络模型对图像进行预测。前面已经介绍了几个非常具有代表性的网络架构，而 TensorFlow 的实现方面可以像前面的章节搭建网络一样，使用 Keras 的 Functional API，但是像 GoogLeNet 或 ResNet 这种较复杂的网络架构，搭建起来还是要花费一点时间的。下面的范例会介绍 Keras Applications 和 TensorFlow Hub 两种搭建 Inception V3 网络架构的方法，并通过两种搭建方法提供的预训练权重来对图像进行预测，如图 9-15 所示。

预测：
- African elephant：80.37%
- Tusker：12.16%
- Indian elephant：0.42%

图 9-15　非洲大象图像经过 Inception V3 网络模型的预测结果

9.2.1 新建项目

建议使用 Jupyter Notebook 来执行本小节的程序代码，操作流程如下：

Step 01 启动 Jupyter Notebook。

在 Terminal（Ubuntu）或命令提示符（Windows）中输入如下指令：

```
jupyter notebook
```

Step 02 新建执行文件。

单击界面右上角的 New 下拉按钮，然后单击所安装的 Python 解释器（在 Jupyter 中都称为 Kernel）来启动它，如图 9-16 所示，显示了 3 个不同的 Kernel：

- Python 3：本地端 Python。
- tf2：虚拟机 Python（TensorFlow-cpu 版本）。
- tf2-gpu：虚拟机 Python（TensorFlow-gpu 版本）。

图 9-16　新建执行文件

Step 03 执行程序代码。

按 [Shift] + [Enter] 快捷键执行单行程序代码，如图 9-17 所示。

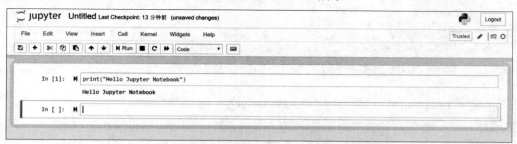

图 9-17　Jupyter 环境界面

接下来，后续的程序代码都可以在 Jupyter Notebook 上执行。

9.2.2　Keras Applications

　　Keras Applications 是提供多种网络模型的高级 API，如表 9-1 所示，可以通过 API 快速搭建网络架构，它提供了已经训练好的网络权重供加载使用。下面的程序以 ResNet50 架构为例。

表 9-1　Keras Applications 提供的网络模型

模型	大小	Top-1 正确率	Top-5 正确率	参数数量	深度
Xception	88 MB	0.790	0.945	22,910,480	126
VGG16	528 MB	0.713	0.901	138,357,544	23
VGG19	549 MB	0.713	0.900	143,667,240	26
ResNet50	99 MB	0.749	0.921	25,636,712	168
InceptionV3	92 MB	0.779	0.937	23,851,784	159
InceptionResNetV2	215 MB	0.803	0.953	55,873,736	572
MobileNet	16 MB	0.704	0.895	4,253,864	88
MobileNetV2	14 MB	0.713	0.901	3,538,984	88
DenseNet121	33 MB	0.750	0.923	8,062,504	121
DenseNet169	57 MB	0.762	0.932	14,307,880	169
DenseNet201	80 MB	0.773	0.936	20,242,984	201
NASNetMobile	23 MB	0.744	0.919	5,326,716	-
NASNetLarge	343 MB	0.825	0.960	88,949,818	-

> **说　明**
>
> Top-1：如果网络预测概率最大的结果与答案相同，就为正确预测，以此标准来计算正确率。
> Top-5：如果网络预测概率最大的 5 个结果，其中一个与答案相同，就为正确预测，以此标准来计算正确率。

Step 01　导入必要的套件。

```
import tensorflow as tf
import numpy as np
```

Step 02　建立 Inception V3 网络架构。

（1）建立 Inception V3 网络架构，并加载 ImageNet 预训练的权重。

```
model = tf.keras.applications.InceptionV3(include_top=True,
                                          weights='imagenet')
```

> **说 明**
>
> **tf.keras.applications.InceptionV3 函数介绍**
>
> 1. include_top：True 为包含全连接层，False 为不包含全连接层。
> 2. weights：加载权重方式，默认为 imagenet。
> (1) None：为 Random Initialization 权重。
> (2) imagenet：为 ImageNet 数据集上预训练网络的权重。
> (3) FILE_PATH：如果输入为模型权重文件的位置，就会直接从文件中读取权重。
> 3. input_tensor（选填）：传入 tf.keras.Input 层。
> 4. input_shape（选填）：设置模型输入大小，默认输入为 (299, 299, 3)，如果需要自定义输入大小，就需要将 include_top 设置为 False。
> 5. pooling（选填）：如果需要自定义 pooling 的方式，就需要将 include_top 设置为 False。
> (1) None（默认）：代表不加入池化层，最后一层输出为卷积层。
> (2) avg：最后一层输出为全局平均池化层（GlobalAveragePooling2D）。
> (3) max：最后一层输出为全局最大池化层（GlobalMaxPooling2D）。
> 6. classes（选填）：输出的类别数量，默认输出类别为 1000，如果需要自定义输出类别，就需要将 include_top 设置为 True，weights 设置为 None。

（2）通过 model.summary 查看网络模型每一层的信息。

```
model.summary()
```

结果如图 9-18 所示。

```
concatenate_1 (Concatenate)     (None, 8, 8, 768)     0         activation_140[0][0]
                                                                activation_141[0][0]
activation_142 (Activation)     (None, 8, 8, 192)     0         batch_normalization_v1_93[0][0]
mixed10 (Concatenate)           (None, 8, 8, 2048)    0         activation_134[0][0]
                                                                mixed9_1[0][0]
                                                                concatenate_1[0][0]
                                                                activation_142[0][0]
avg_pool (GlobalAveragePooling2 (None, 2048)          0         mixed10[0][0]
predictions (Dense)             (None, 1000)          2049000   avg_pool[0][0]
=================================================================
Total params: 23,851,784
Trainable params: 23,817,352
Non-trainable params: 34,432
```

图 9-18 执行结果

（3）将网络模型存储到 TensorBoard 上。

```
model_tb = tf.keras.callbacks.TensorBoard(log_dir='lab8-inceptionv3')
model_tb.set_model(model)
```

（4）启动 TensorBoard（命令行）查看网络模型，如图 9-19 所示。

```
tensorboard --logdir lab8-inceptionv3
```

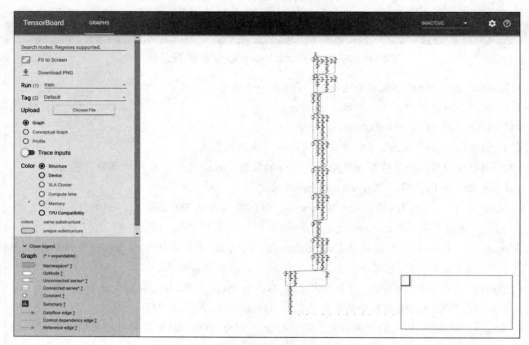

图 9-19　TensorBoard 页面

Step 03 数据预处理和输出解码。

使用 Keras Applications 加载的模型进行预测，对于每个模型 Keras 都提供相对应的数据预处理和输出解码（输出类别）的函数。

- preprocess_input：网络架构的图像预处理。

> **注　意**
>
> 每一个模型在训练时进行数据的归一化并不会相同，例如 VGG、ResNet-50 输入图像为 0~255 的数值，而 inception_v3、xception 输入图像为-1~1 的数值。

- decode_predictions：网络架构的输出解码，即输出预测类别与属于该类别的概率。

导入数据预处理和输出解码的函数：

```
from tensorflow.keras.applications.inception_v3 import preprocess_input
from tensorflow.keras.applications.inception_v3 import decode_predictions
```

Step 04 预测输出结果。

- 建立图像读取的函数

读取图像，并将图像大小缩放至 299×299×3 的尺寸。

```
def read_img(img_path, resize=(299,299)):
    # 读取文件
    img_string = tf.io.read_file(img_path)
```

```
    # 将文件以图像格式来解码
    img_decode = tf.image.decode_image(img_string)
    # 将图像大小调整(resize)到网络输入大小
    img_decode = tf.image.resize(img_decode, resize)
    # 将图像格式增加到4维(batch, height, width, channels),模型预测要求的格式
    img_decode = tf.expand_dims(img_decode, axis=0)
    return img_decode
```

- 从文件夹中读取一张图像(elephant.jpg)用于测试

```
img_path = 'image/elephant.jpg'
# 通过刚建立的函数读取图像
img = read_img(img_path)
# 先将图像转为Integers,再通过matplotlib显示图像
plt.imshow(tf.cast(img, tf.uint8)[0])
```

结果如图 9-20 所示。

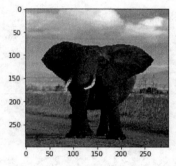

图 9-20　执行结果

- 预测结果

```
img = preprocess_input(img)           # 图像预处理
preds = model.predict(img)            # 预测图像
print("Predicted:", decode_predictions(preds, top=3)[0])#输出预测最高的3个类别
```

结果如下：

```
Predicted: [('n02504458', 'African_elephant', 0.8037859),
            ('n01871265', 'tusker', 0.12163948),
            ('n02504013', 'Indian_elephant', 0.0042992835)]
```

9.2.3　TensorFlow Hub

庞大的深度学习架构通常需要耗费几百甚至几千个小时去训练网络模型，如果能够直接使用开源的网络模型，就可以省去很多不必要的尝试与训练时间。TensorFlow Hub（https://tfhub.dev/，如果这个网站不能直接访问，那么可以访问 https://tensorflow.google.cn/hub）是一个模型共享的平台，类似于 GitHub 或前面介绍的 TensorFlow 数据集，目的是让开发者能够分享彼此的训练模型，如图 9-21 所示。此外，网络模型会根据任务分类，例如语义识别需要的 Embedding 模型或

Classification 模型等。

图 9-21　TensorFlow Hub 网站

Step 01 安装 TensorFlow Hub（命令行）。

```
pip install tensorflow-hub
```

Step 02 导入必要的套件。

```
import tensorflow_hub as hub
```

Step 03 建立 Inception V3 网络架构。

（1）进入 https://tfhub.dev/ 网站，并单击 Classification 按钮，如图 9-22 所示。

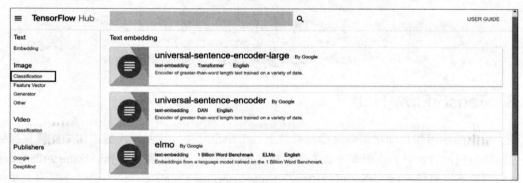

图 9-22　TensorFlow Hub 操作 1

（2）设置 Network 过滤选项为 Inception V3，如图 9-23 所示。

第 9 章　卷积神经网络经典架构 | 219

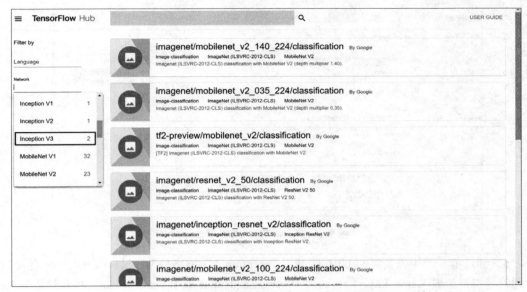

图 9-23　TensorFlow Hub 操作 ②

(3) 单击 TF2 的版本 Inception V3 模型，如图 9-24 所示。

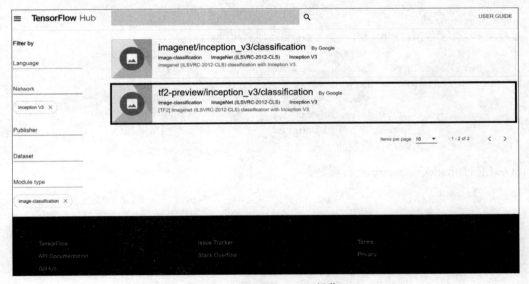

图 9-24　TensorFlow Hub 操作 3

（4）页面中会显示模型的信息及使用方式，然后单击 Copy URL 按钮，如图 9-25 所示。

说　明
在 TensorFlow Hub 模型信息页面提供了使用说明的介绍，一般都会提供数据集的标签文件（ImageNetLabels.txt）和数据的输入格式（Input Size：299×299×3，数值范围：0~1），在后面的步骤会用到一些信息来建立数据预处理和输出解码器。

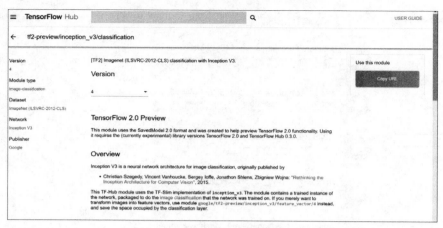

图 9-25　TensorFlow Hub 操作 4

（5）通过前一个步骤复制的 URL 来建立 Inception V3 网络模型：hub.KerasLayer 会将 URL 指引到的模型封装成 Keras Layer，再通过 tf.keras.Sequential 或 tf.keras.Model，将其封装为 Keras Model。

```
# Inception V3 预训练模型的 URL
module_url = "https://tfhub.dev/google/tf2-preview/inception_v3/classification/4"
# 建立一个序贯模型，网络模型中包含 Inception V3 网络层
model = tf.keras.Sequential([
    # hub.KerasLayer 将加载的 Inception V3 模型封装成网络层
    hub.KerasLayer(module_url,
                input_shape=(299, 299, 3),      # 模型输入大小
                output_shape=(1001, ),           # 模型输出大小
                name='Inception_v3')             # 网络层名称
])
```

（6）通过 model.summary 查看模型。

```
model.summary()
```

结果如图 9-26 所示。

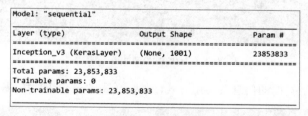

图 9-26　执行结果

Step 04　数据预处理和输出解码。

- 建立数据预处理函数

```
def read_img(img_path, resize=(299,299)):
```

```
    # 读取文件
    img_string = tf.io.read_file(img_path)
    # 将读入文件以图像格式来解码
    img_decode = tf.image.decode_image(img_string)
    # 图像大小调整，将其大小调整成 Inception V3 的输入尺寸
    img_decode = tf.image.resize(img_decode, resize)
    # 图像归一化，缩放到 0~1
    img_decode = img_decode / 255.0
    # 网络预测需增加一个维度
    img_decode = tf.expand_dims(img_decode, axis=0)
    return img_decode
```

- 建立输出解码器

```
# 下载 ImageNet 的标签文件，标签文件网址可以在 TensorFlow Hub 模型信息页面找到
labels_path=tf.keras.utils.get_file('ImageNetLabels.txt', 'https://storage.googleapis.com/download.tensorflow.org/data/ImageNetLabels.txt')
# 读取标签文件中的数据
with open(labels_path) as file:
    lines = file.read().splitlines()
# 显示读取的标签
print(lines)
# 将标签转成 NumPy array，作为网络输出的解码器
imagenet_labels = np.array(lines)
```

结果如图 9-27 所示。

```
['background', 'tench', 'goldfish', 'great white shark', 'tiger shark', 'hammerhead', 'electric ray', 'stingray', 'cock',
'hen', 'ostrich', 'brambling', 'goldfinch', 'house finch', 'junco', 'indigo bunting', 'robin', 'bulbul', 'jay', 'magpie',
'chickadee', 'water ouzel', 'kite', 'bald eagle', 'vulture', 'great grey owl', 'European fire salamander', 'common newt',
'eft', 'spotted salamander', 'axolotl', 'bullfrog', 'tree frog', 'tailed frog', 'loggerhead', 'leatherback turtle', 'mud
turtle', 'terrapin', 'box turtle', 'banded gecko', 'common iguana', 'American chameleon', 'whiptail', 'agama', 'frilled l
izard', 'alligator lizard', 'Gila monster', 'green lizard', 'African chameleon', 'Komodo dragon', 'African crocodile', 'A
merican alligator', 'triceratops', 'thunder snake', 'ringneck snake', 'hognose snake', 'green snake', 'king snake', 'gart
er snake', 'water snake', 'vine snake', 'night snake', 'boa constrictor', 'rock python', 'Indian cobra', 'green mamba',
'sea snake', 'horned viper', 'diamondback', 'sidewinder', 'trilobite', 'harvestman', 'scorpion', 'black and gold garden s
pider', 'barn spider', 'garden spider', 'black widow', 'tarantula', 'wolf spider', 'tick', 'centipede', 'black grouse',
'ptarmigan', 'ruffed grouse', 'prairie chicken', 'peacock', 'quail', 'partridge', 'African grey', 'macaw', 'sulphur-crest
ed cockatoo', 'lorikeet', 'coucal', 'bee eater', 'hornbill', 'hummingbird', 'jacamar', 'toucan', 'drake', 'red-breasted m
erganser', 'goose', 'black swan', 'tusker', 'echidna', 'platypus', 'wallaby', 'koala', 'wombat', 'jellyfish', 'sea anemon
e', 'brain coral', 'flatworm', 'nematode', 'conch', 'snail', 'slug', 'sea slug', 'chiton', 'chambered nautilus', 'Dungene
ss crab', 'rock crab', 'fiddler crab', 'king crab', 'American lobster', 'spiny lobster', 'crayfish', 'hermit crab', 'isop
od', 'white stork', 'black stork', 'spoonbill', 'flamingo', 'little blue heron', 'American egret', 'bittern', 'crane', 'l
impkin', 'European gallinule', 'American coot', 'bustard', 'ruddy turnstone', 'red-backed sandpiper', 'redshank', 'dowitc
her', 'oystercatcher', 'pelican', 'king penguin', 'albatross', 'grey whale', 'killer whale', 'dugong', 'sea lion', 'Chihu
ahua', 'Japanese spaniel', 'Maltese dog', 'Pekinese', 'Shih-Tzu', 'Blenheim spaniel', 'papillon', 'toy terrier', 'Rhodesi
```

图 5-27 执行结果

Step 05 预测输出结果。

- 从文件夹中读取一张图像（elephant.jpg）用于测试

```
img_path = 'image/elephant.jpg'
# 通过刚建立的函数读取图像
img = read_img(img_path)
# 通过 matplotlib 显示图像
plt.imshow(img[0])
```

结果如图 9-28 所示。

图 9-28 执行结果

- 预测结果

```
preds = model.predict(img)         # 预测图像
index = np.argmax(preds)           # 获取预测结果最大的 index
print("Predicted:", imagenet_labels[index])   # 通过解码器将输出转成标签
```

结果如下：

```
Predicted: African elephant
```

- 显示最好的 3 个预测结果

```
# 获取预测结果最大的 3 个 indexs
top3_indexs = np.argsort(preds)[0, ::-1][:3]
# 通过解码器将输出转成标签
print("Predicted:", imagenet_labels[top3_indexs])
```

结果如下：

```
Predicted: ['African elephant' 'tusker' 'Indian elephant']
```

9.3 参考文献

[1] Lecun Y, Bottou L, Bengio Y, et al. Gradient-based learning applied to document recognition [J]. in Proceedings of the IEEE, 1998. 86(11):2278-2324.

[2] Krizhevsky A, Sutskever I, Hinton G E. Imagenet classification with deep convolutional neural networks[C]. Advances in Neural Information Processing Systems, 2012, 1097-1105.

[3] Simonyan K, Zisserman A. Very deep convolutional networks for large-scale image recognition [C]. In International Conference on Learning Representations, 2015, 1-14.

[4] Szegedy C. Going deeper with convolutions [C]. In IEEE Conference on Computer Vision and Pattern Recognition, 2015, 1-9.

[5] Ioffe S, Szegedy C. Batch normalization: Accelerating deep network training by reducing internal covariate shift [C]. In International Conference on Machine Learning, 2015, 448-456.

[6] Szegedy C, Vanhoucke V, Ioffe S, et al. Rethinking the inception architecture for computer vision [C]. Proceedings of the IEEE Conference on Computer Vision and Pattern Recognition, 2016.

[7] Szegedy C, Ioffe S, Vanhoucke V, et al. Inception-V4, inception-ResNet and the impact of residual connections on learning [C]. In Association for the Advancement of Artificial Intelligence, 2017, 1-3.

[8] Chollet F. Xception: Deep learning with depthwise separable convolutions [C]. Proceedings of the IEEE Conference on Computer Vision and Pattern Recognition, 2017.

[9] He K, Zhang X, Ren S, et al. Deep residual learning for image recognition[C]. Proceedings of the IEEE Conference on Computer Vision and Pattern Recognition, 2016, 770-778.

第 10 章

迁移学习

学习目标

- 认识迁移学习的好处
- 介绍迁移学习的训练技巧
- 通过 TensorFlow Hub 搭建 Inception V3 网络模型来进行迁移学习

10.1 迁移学习

10.1.1 迁移学习介绍

5.1.3 小节说明了不同深度的卷积层能够识别的特征会有所不同。简而言之，前面几层卷积层主要负责识别边缘或线条等，后面几层卷积层基于前方提取到的特征，能够识别更为具体的特征，例如鼻子、眼睛或耳朵等，如图 10-1 所示。

这些学习到的特征能够应用在不同的任务上，举一个经典的例子：使用拥有数万笔训练数据的 ImageNet 数据集训练网络模型，大量的数据使网络模型得以学习到非常多样的特征，而这种网络模型称为预训练模型（Pre-Trained Model）。当我们有一个新的数据集要训练时，可以不必搭建一个全新的网络模型从头训练，而将训练构建在已有优秀特征提取能力的预训练模型上。这种训练方式被称为迁移学习（Transfer Learning）[1]-[9]。

迁移学习其实很像我们人类的学习方式，通过完成某个任务所获得的知识同样可用于解决其他相关的任务。根据如今许多重要文献的实验结果，可以发现图像任务上使用 ImageNet 的预训练网络迁移学习比从头开始训练模型的性能要好，且训练时间更短。

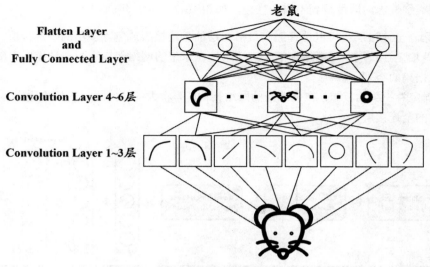

图 10-1　深层卷积网络

10.1.2　迁移学习训练技巧

迁移学习的训练主要根据两种情况，且大致分成 4 种不同的训练方法。

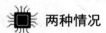 两种情况

两种情况分别为：

- 新数据集的大小

如果新数据集的数据量为几万笔数据，就属于大数据集；如果新数据集的数据量为几千或几百笔数据，就属于小数据集。

- 数据集的相似程度

新数据集与预训练模型用的数据集之间的相似程度。例如，猫与老虎属于相似度高的数据，而猫与桌子则属于相似度低的数据。

 4 种不同的训练方法

4 种不同的训练方法分别为：

- 小数据集、相似数据

小数据集在庞大的网络架构上训练时容易发生过拟合问题，因此预训练模式的权重必须保持不变。由于新数据集与预训练模式使用的数据集相似性高，新数据集在每一层卷积层都有相似的特征，尤其是更高层的卷积层，因此提取特征的卷积层不需要改变，只对处理特征分类的全连接层进行改变即可。故而我们将最后几层全连接层删除，并加上新的全连接层，再重新将这些相似的特征

连接到新的类别输出,即可得到不错的性能,如图10-2所示。步骤如下:

(1)删除全连接层:可以选择删除最后一层的全连接层或删除多层的全连接层。

(2)新增全连接层:将新增加的全连接层接在原来的网络架构后面,且最后一层全连接层的输出与新数据集的类别一样。

(3)固定卷积层的权重:在训练新的网络架构时,将大部分网络层的权重固定,不进行训练,只训练新增的全连接层。

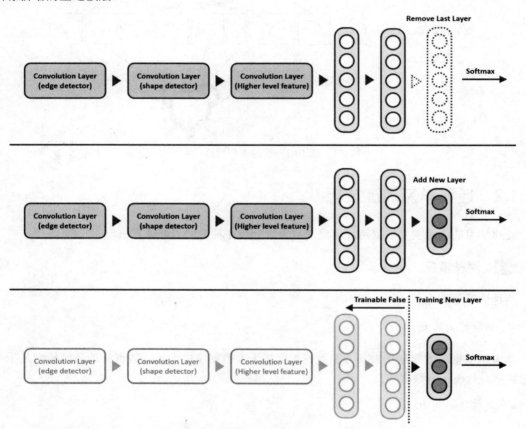

图 10-2 小数据集、相似数据的训练方法

- **小数据集、不相似数据**

因为小数据集在庞大的网络架构上训练时容易发生过拟合问题,所以预训练模型的权重必须保持不变。由于新数据集与预训练模型使用的数据集相似度低,新数据集只有在低层卷积层有相似特征,更高层卷积层的特征大多不相似,因此只需要保留低层卷积层,其他卷积层和全连接层都删除,并加入新的全连接层,如图10-3所示。步骤如下:

(1)删除全连接层:删除大部分网络,只保留前面小部分网络层(提取线条、颜色或纹路的网络层)。

(2)新增全连接层:将新增加的全连接层接在原来的网络架构后面,且最后一层全连接层输出与新数据集的类别一样。

（3）固定卷积层的权重：在训练新网络架构时，将大部分网络层的权重固定，不进行训练，只训练新增的全连接层。

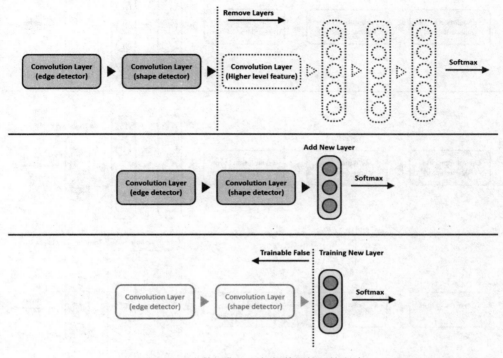

图 10-3 小数据集、不相似数据的训练方法

- 大数据集、相似数据

拥有大量的数据集一般不需要担心在训练过程中发生过拟合问题，因此可以对整个预训练模型或全新的随机初始化权重的网络模型进行训练。下面提供了两种训练方法：

方法①：由于新数据集与预训练模型使用的数据集相似度高，新数据集在每一层卷积层都有相似的特征，尤其是更高层的卷积层，因此提取特征的卷积层不必改变，只改变处理特征分类的全连接层即可。所以，我们将最后几层全连接层删除，并加上新的全连接层，再重新将这些相似的特征连接到新的类别输出，即可得到不错的性能。最后，因为数据集够大，我们可以尝试将整个网络模型进行微调，让网络层提取到的特征贴近新的数据，也许网络模型会得到更好的性能，如图 10-4 所示。

（1）删除全连接层：可以选择删除最后一层的全连接层或删除多层的全连接层。

（2）新增全连接层：将新增加的全连接层接在原来的网络架构后面，且最后一层全连接层输出与新数据集的类别一样。

（3）固定卷积层的权重：在训练新的网络架构时，将大部分网络层的权重固定，不进行训练，只训练新增的全连接层。

（4）训练整个网络模型：该步骤可选择省略，通过再一次微调整个网络模型有可能会提升网络模型的性能。

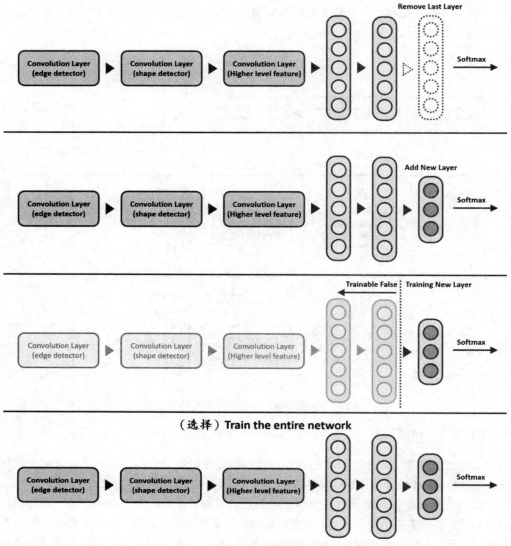

图 10-4　大数据集、相似数据的训练方法 1

方法 2：一般网络的训练方法，因为数据集够大，所以可以直接建立一个全新的随机初始化权重的网络模型，并重新训练整个网络，如图 10-5 所示。

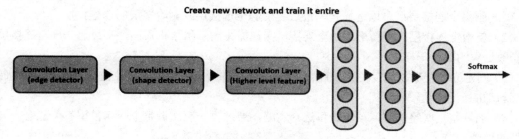

图 10-5　大数据集、相似数据的训练方法 2

- 大数据集、不相似数据

拥有大量的数据集一般不需要担心在训练过程中发生过拟合问题，因此可以对整个预训练模型或全新的随机初始化权重的网络模型进行训练，下面提供了两种训练方法：

方法 1：由于新数据集与预训练模型使用的数据集相似度低，新数据集与大多数网络层的特征都不相似，因此我们先将最后几层全连接层删除，并加上新的全连接层，再通过对整个预训练模型进行训练，以校正特征的偏差，如图 10-6 所示。

（1）删除全连接层：可以选择删除最后一层的全连接层或删除多层的全连接层。
（2）新增全连接层：将新增加的全连接层接在原来的网络架构后面，且最后一层全连接层输出与新数据集的类别一样。
（3）训练整个网络模型：重新微调整个网络模型。

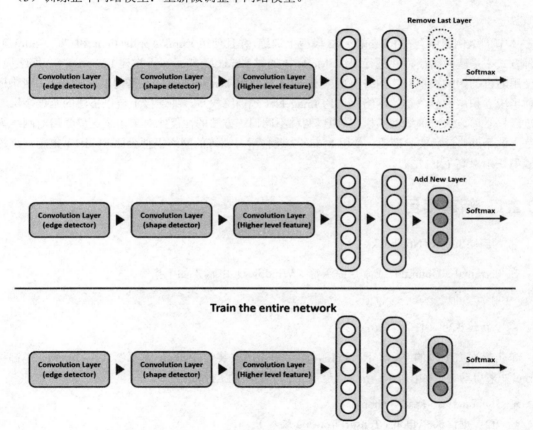

图 10-6　大数据集、不相似数据的训练方法 1

方法 2：一般网络的训练方法，因为数据集够大，所以可以直接建立一个全新的随机初始化权重的网络模型，并重新训练整个网络，如图 10-7 所示。

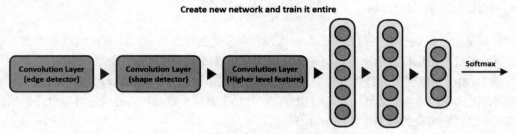

图 10-7 大数据集、不相似数据的训练方法 2

10.2 实验：迁移学习范例

本节的范例程序会在 Inception V3 模型上训练，并且使用 Keras Applications 和 TensorFlow Hub 两种方式来搭建网络模型并进行训练。Model-1 通过 Keras Applications 搭建 Inception V3 网络模型，并使用随机初始的权重训练网络模型；Model-2 则使用 TensorFlow Hub 搭建 Inception V3 网络模型的前半段卷积层部分，初始网络权重为 ImageNet 预训练权重，并连接上自己搭建的全连接层，再通过迁移学习方法来训练网络模型。通过两种不同训练方法的性能比对来了解迁移学习对网络模型训练上的帮助。最后实验证明，使用 Model 2 迁移学习训练的网络模型会比 Model 1 重新训练的网络模型正确率高 1.4%左右。

10.2.1 新建项目

Step 01 启动 Jupyter Notebook。

在 Terminal（Ubuntu）或命令提示符（Windows）中输入如下指令：

```
jupyter notebook
```

Step 02 新建执行文件。

单击界面右上角的 New 下拉按钮，然后单击所安装的 Python 解释器（在 Jupyter 中都称为 Kernel）来启动它，如图 10-8 所示，显示了 3 个不同的 Kernel：

- Python 3：本地端 Python。
- tf2：虚拟机 Python（TensorFlow-cpu 版本）。
- tf2-gpu：虚拟机 Python（TensorFlow-gpu 版本）。

第 10 章 迁移学习

图 10-8 新建执行文件

Step 03 执行程序代码。

按 Shift + Enter 快捷键执行单行程序代码，如图 10-9 所示。

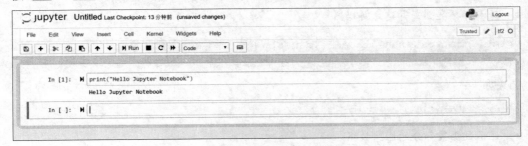

图 10-9 Jupyter 环境界面

接下来，后续的程序代码都可以在 Jupyter Notebook 上执行。

10.2.2 数据集介绍

本节使用的是 Cats vs Dogs 数据集，由微软研究团队提供，总共为 23 262 张猫与狗的照片，并且将训练数据、验证数据与测试数据在数据集所占的比例设置为 8:1:1。本节的任务只需分出猫和狗两个类别，所以是二分类问题。

10.2.3 程序代码

Step 01 导入必要的套件。

```
import os
import numpy as np
import tensorflow as tf
import tensorflow_hub as hub
import tensorflow_datasets as tfds
from tensorflow import keras
from tensorflow.keras import layers
import matplotlib.pyplot as plt
# 从文件夹的 preprocessing.py 文件中导入 flip、color、rotate、zoom 图像增强函数
```

```
from preprocessing import flip, color, rotate, zoom
```

Step 02 读取数据并分析。

- 加载 cats_vs_dogs 数据集

```
# 将训练数据重新分成 8:1:1 等分，分别分给训练数据、验证数据和测试数据
train_split, valid_split, test_split = tfds.Split.TRAIN.subsplit([80, 10, 10])
# 获取训练数据，并读取 data 的信息
train_data, info = tfds.load("cats_vs_dogs", split=train_split,
                              with_info=True)
# 获取验证数据
valid_data = tfds.load("cats_vs_dogs", split=valid_split)
# 获取测试数据
test_data = tfds.load("cats_vs_dogs", split=test_split)
```

- 查看标签类别，并建立解码器

```
print(info.features['label'].names)            # 显示数据集类别
decoder = info.features['label'].names         # 建立解码器
```

结果如下：

```
['cat', 'dog']
```

- 显示数据集部分的图像数据

```
for data in train_data.take(1):
    img = data['image']        # 读取图像
    label = data['label']      # 读取标签
# 通过解码器获取类别名称
plt.title(decoder[label])
# 显示图像
plt.imshow(img)
```

结果如图 10-10 所示。

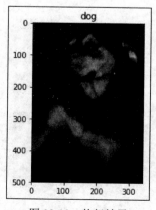

图 10-10　执行结果

Step 03 数据集设置。

数据预处理：

- 训练数据
  - 归一化：将图像中的所有像素除以 255，也就是将像素值缩放到 0~1。
  - 调整图像的大小：将输入图像调整为网络模型要求的格式（299, 299, 3）。
  - 图像增强：将图像进行水平翻转、旋转、缩放和颜色转换。
- 验证和测试数据
  - 归一化：将图像中的所有像素除以 255，也就是将像素值缩放到 0~1。
  - 调整图像的大小：将输入图像调整为网络模型要求的格式（299, 299, 3）。

```python
input_shape = (299, 299)  # 设置网络输入的大小

def parse_aug_fn(dataset):
    """
    Image Augmentation（图像增强）函数
    """
    # 图像归一化
    x = tf.cast(dataset['image'], tf.float32) / 255.
    x = tf.image.resize(x, input_shape)
    # 随机水平翻转
    x = flip(x)
    # 触发颜色转换概率 50%
    x = tf.cond(tf.random.uniform([], 0, 1) > 0.5, lambda:color(x), lambda: x)
    # 触发图像旋转概率 0.25%
    x = tf.cond(tf.random.uniform([], 0, 1) > 0.75, lambda:rotate(x), lambda: x)
    # 触发图像缩放概率 50%
    x = tf.cond(tf.random.uniform([], 0, 1) > 0.5, lambda: zoom(x), lambda: x)
    return x, dataset['label']

def parse_fn(dataset):
    # 图像归一化
    x = tf.cast(dataset['image'], tf.float32) / 255.
    x = tf.image.resize(x, input_shape)
    return x, dataset['label']
```

数据集设置：

```python
AUTOTUNE = tf.data.experimental.AUTOTUNE    # 自动调整模式
buffer_size = 1000    # 因为这次的图像较大，缓存空间设置为 1000
batch_size = 64       # 批量大小

# 加载预处理 parse_aug_fn 函数，CPU 数量为自动调整模式
train_data = train_data.map(map_func=parse_aug_fn,
                            num_parallel_calls=AUTOTUNE)
# 打乱数据集
train_data = train_data.shuffle(train_num)
# 设置批量大小，并启用 prefetch 模式（缓存空间为自动调整模式）
train_data = train_data.batch(batch_size).prefetch(buffer_size=AUTOTUNE)
```

```
# 加载预处理 parse_fn 函数，CPU 数量为自动调整模式
valid_data = valid_data.map(map_func=parse_fn, num_parallel_calls=AUTOTUNE)
# 设置批量大小，并启用 prefetch 模式（缓存空间为自动调整模式）
valid_data = valid_data.batch(batch_size).prefetch(buffer_size=AUTOTUNE)

# 加载预处理 parse_fn 函数，CPU 数量为自动调整模式
test_data = test_data.map(map_func=parse_fn, num_parallel_calls=AUTOTUNE)
# 设置批量大小，并启用 prefetch 模式（缓存空间为自动调整模式）
test_data = test_data.batch(batch_size).prefetch(buffer_size=AUTOTUNE)
```

Step 04 训练 Model-1。

- 建立模型存储位置

```
model_dir = 'lab9-logs/models'        # 设置存储权重的目录
os.makedirs(model_dir)                # 创建存储权重的目录
```

- 设置回调函数

```
# 存储训练记录文件
log_dir = os.path.join('lab9-logs', 'model-1')
model_cbk = keras.callbacks.TensorBoard(log_dir=log_dir)
# 设置停止训练的条件（当正确率超过 30 次迭代没有上升时，训练会终止）
model_esp = keras.callbacks.EarlyStopping(monitor='val_binary_accuracy',
                                          patience=30,
                                          mode='max')
```

- 建立 Inception V3 网络模型

```
# 建立模型（不包含全连接层和预训练权重），最后一层卷积加上 GlobalAveragePooling
base_model = tf.keras.applications.InceptionV3(include_top=False,
                                               weights=None,
                                               pooling='avg',
                                               input_shape=input_shape+(3,))
# 将刚建立的 Inception V3 模型接上两层全连接层，并且最后一层使用 Sigmoid 输出
model_1 = tf.keras.Sequential([
    base_model,
    # 最后接上两层全连接层，并且输出使用 Sigmoid
    layers.Dense(128, activation='relu'),
    layers.Dense(1, activation='sigmoid')
])
```

- 通过 model.summary 查看网络模型信息

```
model_1.summary()
```

结果如图 10-11 所示。

```
Model: "sequential"
_____
Layer (type)                 Output Shape              Param #
=================================================================
inception_v3 (Model)         (None, 2048)              21802784
_____
dense (Dense)                (None, 128)               262272
_____
dense_1 (Dense)              (None, 1)                 129
=================================================================
Total params: 22,065,185
Trainable params: 22,030,753
Non-trainable params: 34,432
```

图 10-11　执行结果

- 设置训练使用的优化器、损失函数和评价指标函数

```
model_1.compile(keras.optimizers.Adam(),
            loss=keras.losses.BinaryCrossentropy(),
            metrics=[keras.metrics.BinaryAccuracy()])
```

- 训练网络模型

```
history = model_1.fit(train_data,
            epochs=200,
            validation_data=valid_data,
            callbacks=[model_cbk, model_esp])
```

结果如图 10-12 所示。

```
Epoch 86/200
582/582 [==============================] - 290s 499ms/step - loss: 0.0359 - binary_accuracy: 0.9864 - val_loss: 0.0499 - val_binary_accuracy: 0.9793
Epoch 87/200
582/582 [==============================] - 291s 499ms/step - loss: 0.0326 - binary_accuracy: 0.9880 - val_loss: 0.0673 - val_binary_accuracy: 0.9750
Epoch 88/200
582/582 [==============================] - 291s 499ms/step - loss: 0.0386 - binary_accuracy: 0.9858 - val_loss: 0.0633 - val_binary_accuracy: 0.9767
Epoch 89/200
582/582 [==============================] - 290s 498ms/step - loss: 0.0346 - binary_accuracy: 0.9865 - val_loss: 0.0738 - val_binary_accuracy: 0.9707
Epoch 90/200
582/582 [==============================] - 290s 498ms/step - loss: 0.0361 - binary_accuracy: 0.9858 - val_loss: 0.0678 - val_binary_accuracy: 0.9759
Epoch 91/200
582/582 [==============================] - 290s 498ms/step - loss: 0.0347 - binary_accuracy: 0.9868 - val_loss: 0.0744 - val_binary_accuracy: 0.9728
```

图 10-12　执行结果

Step 05 训练 Model-2（迁移学习）。

- 设置回调函数

```
# 存储训练记录文件
log_dir = os.path.join('lab9-logs', 'model-2')
model_cbk = keras.callbacks.TensorBoard(log_dir=log_dir)
# 设置停止训练的条件（当正确率超过 30 次迭代没有上升时，训练会终止）
model_esp = keras.callbacks.EarlyStopping(monitor='val_binary_accuracy',
                                          patience=30,
                                          mode='max')
```

- 建立 Inception V3 网络模型

```
# Inception V3 预训练模型的 URL（模型不包含全连接层）
module_url = "https://tfhub.dev/google/tf2-preview/inception_v3/feature_vector/2"
# 建立一个序贯模型，网络模型包含 Inception V3 网络层和两层全连接层
model_2 = tf.keras.Sequential([
    # hub.KerasLayer 将加载 Inception V3 模型的前半段，并封装成网络层
    hub.KerasLayer(module_url,
                   input_shape=(299, 299, 3),  # 模型输入大小
                   output_shape=(2048,),        # 模型输出大小
                   trainable=False),            # 将模型训练权重设置为 False（冻结）
    # 最后接上两层全连接层，并且输出使用 Sigmoid
    layers.Dense(128, activation='relu'),
    layers.Dense(1, activation='sigmoid')
])
```

- 通过 model.summary 查看网络模型信息

```
model_2.summary()
```

结果如图 10-13 所示。

```
Model: "sequential_1"
_____
Layer (type)                 Output Shape              Param #
=================================================================
keras_layer (KerasLayer)     (None, 2048)              21802784
_____
dense_2 (Dense)              (None, 128)               262272
_____
dense_3 (Dense)              (None, 1)                 129
=================================================================
Total params: 22,065,185
Trainable params: 262,401
Non-trainable params: 21,802,784
```

图 10-13　执行结果

- 设置训练使用的优化器、损失函数和评价指标函数

```
model_2.compile(keras.optimizers.Adam(),
                loss=keras.losses.BinaryCrossentropy(),
                metrics=[keras.metrics.BinaryAccuracy()])
```

- 训练网络模型

```
history = model_2.fit(train_data,
                      epochs=200,
                      validation_data=valid_data,
                      callbacks=[model_cbk, model_esp])
```

结果如图 10-14 所示。

```
Epoch 20/200
582/582 [==============================] - 90s 155ms/step - loss: 0.0339 - binary_accuracy: 0.9864 - val_loss: 0.0180 - v
al_binary_accuracy: 0.9927
Epoch 21/200
582/582 [==============================] - 89s 154ms/step - loss: 0.0341 - binary_accuracy: 0.9864 - val_loss: 0.0184 - v
al_binary_accuracy: 0.9922
Epoch 22/200
582/582 [==============================] - 90s 154ms/step - loss: 0.0330 - binary_accuracy: 0.9867 - val_loss: 0.0204 - v
al_binary_accuracy: 0.9922
Epoch 23/200
582/582 [==============================] - 90s 155ms/step - loss: 0.0333 - binary_accuracy: 0.9879 - val_loss: 0.0205 - v
al_binary_accuracy: 0.9901
Epoch 24/200
582/582 [==============================] - 89s 154ms/step - loss: 0.0330 - binary_accuracy: 0.9868 - val_loss: 0.0192 - v
al_binary_accuracy: 0.9914
Epoch 25/200
582/582 [==============================] - 90s 154ms/step - loss: 0.0327 - binary_accuracy: 0.9869 - val_loss: 0.0236 - v
al_binary_accuracy: 0.9905
```

图 10-14 执行结果

说 明

大家可能注意到了，在第 8 章通过 TensorFlow Hub 所加载的 Inception V3 URL 为 https://tfhub.dev/google/tf2-preview/inception_v3/classification/2，而本章为 https://tfhub.dev/google/tf2-preview/inception_v3/feature_vector/2。差别在于：前者加载的 Inception V3 模型包含最后一层分类层（1000 个类别），而后者删除了最后一层分类层，直接输出 2048 维的特征。

Step 05 测试 Model-1 和 Model-2 的训练结果。

从结果来看，迁移学习的训练方法（Model 2）比从头训练的训练方法正确率高 1.4%左右，且训练速度快了许多。

```
# 加载 Model 1 最佳正确率的权重
model_1.load_weights(model_dir + '/Best-model-1.h5')
# 加载 Model 2 最佳正确率的权重
model_2.load_weights(model_dir + '/Best-model-2.h5')
# 计算 Model 1 和 Mode12 的损失值和正确率
loss_1, acc_1 = model_1.evaluate(test_data)
loss_2, acc_2 = model_2.evaluate(test_data)
print("Model_1 Prediction: {}%".format(acc_1 * 100))
print("Model_2 Prediction: {}%".format(acc_2 * 100))
```

结果如下：

```
Model_1 Prediction: 97.97413945198059%
Model_2 Prediction: 99.39655065536499%
```

10.3　参考文献

[1] Yosinski J, Clune J, Bengio Y, et al. How transferable are features in deep neural networks[C]. Advances in Neural Information Processing Systems, pages, 2014, 3320-3328.

[2] Oquab M, Bottou L, Laptev I, et al. Learning and Transferring Mid-level Image Representations Using Convolutional Neural Networks [C]. Proceedings of the IEEE Conference on Computer Vision and Pattern Recognition, Columbus, 2014, 1717-1724.

[3] Mormont R, Geurts P, Marée R. Comparison of Deep Transfer Learning Strategies for Digital Pathology[C]. Proceedings of the IEEE Conference on Computer Vision and Pattern Recognition Workshops, 2018, 2343-2343.

[4] Kornblith S, Shlens J, Le Q V. Do Better ImageNet Models Transfer Better[C]. Proceedings of the IEEE Conference on Computer Vision and Pattern Recognition, 2019, 2661-2671.

[5] Shin H. Deep Convolutional Neural Networks for Computer-Aided Detection: CNN Architectures, Dataset Characteristics and Transfer Learning [J]. In IEEE Transactions on Medical Imaging, 2016, 35(5):1285-1298.

[6] Tajbakhsh N. Convolutional Neural Networks for Medical Image Analysis: Full Training or Fine Tuning[J]. In IEEE Transactions on Medical Imaging, 2016, 35(5):1299-1312.

[7] Ganin Y, Ustinova E, Ajakan H, et al. Domain-adversarial training of neural networks [J]. In The Journal of Machine Learning Research, 2016, 17(1): 2096-2030.

[8] Hendrycks D, Lee K, Mazeika M. Using Pre-Training Can Improve Model Robustness and Uncertainty [C]. In International Conference on Machine Learning, 2019.

[9] Ding Z, Fu Y. Deep Transfer Low-Rank Coding for Cross-Domain Learning [J]. In IEEE Transactions on Neural Networks and Learning Systems, 2019, 30(6): 1768-1779.

第 11 章

变分自编码器

学习目标

- 介绍自编码器的网络架构
- 介绍变分自编码器的网络架构
- 实现变分自编码器项目，生成手写数字

11.1 自编码器介绍

在进入变分自编码器（Variational Auto-Encoder，VAE）主题之前，先介绍它的前身自编码器（Auto-Encoder，AE）[1]-[3]。自编码器是一种无监督式学习的算法，网络架构主要分为两部分：编码器（Encoder）和解码器（Decoder）。编码器负责将图像信息压缩到任意维度的低维隐空间，即低维隐空间（Low-Dimensional Latent Space），低维空间的压缩信息简称编码（Code），而解码器则是将编码解码回原来的图像。下面分成两部分来解释，分别为自编码器训练和自编码器生成图像。

自编码器训练机制

自编码器训练的目标是希望输入编码器的图像与解码器输出的图像越相似越好，例如 MNIST 手写数字图像，通过编码器将输入（Input）28×28 的图像压缩成二维向量的编码，再经由解码将二维向量的编码解码回原来 28×28 的图像（Output），而神经网络模型的训练目标是希望输出图像与输入图像越相似越好，如图 11-1 所示。整个训练过程只需要图像信息，不需要任何标签，因此被称为无监督式学习。

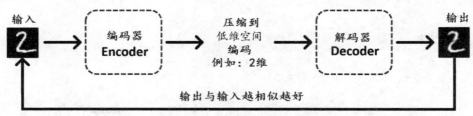

图 11-1 自编码器训练示意图

自编码器生成图像

当训练好自编码器网络模型后,我们可以随机产生一组编码,并将编码输入解码器中,希望解码器能够产生一组与训练数据类似且有意义的数据,如图 11-2 所示。

图 11-2 解码器生成图像

但是,在训练自编码器时,并没有对编码器产生编码的分布进行约束,所以随机生成的编码通过解码器生成的图像通常不具有任何意义,且相似的编码输出之间没有连续关系。例如,生成 225 组二维向量的编码,编码的值是从-1.5 到+1.5 线性采样生成的,而这 225 组编码经过解码器解码生成 225 张图像,并非每一张图像都可以还原成与原数据相似的图像,如图 11-3 所示,除了右下角可以有效生成 3、5、8 图像之外,其余部分皆无法生成有效的图像。

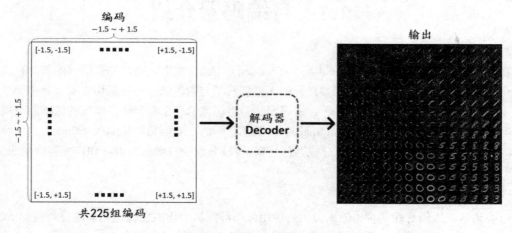

图 11-3 从自编码器输入的 225 组编码经过解码器解码生成 225 张图像

11.2 变分自编码器介绍

变分自编码器[4]是自编码器的高级版，网络模型是由编码器和解码器两部分组成的，它也是无监督式学习。不过，与自编码器不同，变分自编码器的编码器网络架构在训练后会输出两个向量，分别为均值（Mean，μ）和方差（Variance，$σ^2$），再由均值和方差来产生正态分布，而解码器的输入就是从编码器所生成的正态分布中随机抽取一个点作为输入编码，如图11-4所示。

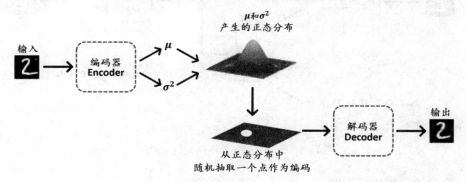

图 11-4 变分自编码器的概念图

至于变分自编码器为什么将编码设计为正态分布，这里使用较直观的例子来解释。如图 11-5 所示，输入两张数字 1 的图像 Input A 和 Input B，经过编码器会产生编码 A 和编码 B 的正态分布，从分布中抽取的数值经由解码器还原回输入图像相似的输出图像 Output A 和 Output B，两个分布之间会有交集，而交集的部分可能会产生介于 Output A 和 Output B 之间的图像，也就是图中的输出图像 Output C。这样的设计使得输出之间有连续关系，像是 Output A、Output B 和 Output C 之间的连续关系。

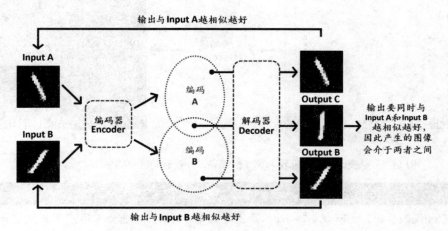

图 11-5 正态分布编码的优点

但是变分自编码器在实现上并不能像图 11-4 那样，因为从正态分布中随机抽取样本（采样或抽样），采样这个操作是不可导的，反向传递不会有数值。因此，我们将从正态分布抽取一个编码（Code）改为从标准正态分布抽取一个 e，然后让 Code=exp(σ^2)×e+μ，如图 11-6 所示。另一个更直观的解释：μ 为原来自编码器的 Code，而这个 Code 会去加一个 Noise（exp(σ^2)×e），希望加上 Noise（噪音）的编码仍然可以解码还原到原图，而 Noise 是从一个标准正态分布产生的值，并且乘上一个放大倍率 exp(σ^2)。

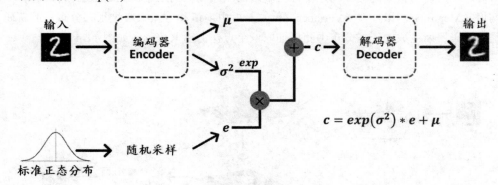

图 11-6　变分自编码器网络架构示意图

变分编码器生成图像

与图 11-3 使用相同的 225 组编码，经过变分自编码器的解码器解码产生 225 张图像，如图 11-7 所示。而这 225 张图像之间是有相关性的，例如从左到右数字可能会慢慢向右倾斜，而从上到下数字可能会慢慢出现圆圈的特征，这都是因为加入了噪音的机制。

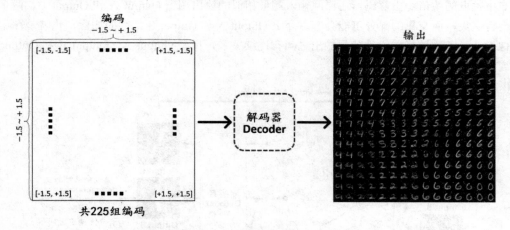

图 11-7　从变分自编码器输入 225 组编码，经过解码器解码生成 225 张图像

说　明
方差（Variance，σ^2）之所以要取 exp，是因为编码器的输出可能为负值，所以加上 exp 可以让输出恒为正值。

11.3 变分自解码器的损失函数

变分自解码器的训练目标是希望预测输出的图像与输入的图像越相似越好，因此对输入图像与输出图像的每一个像素都做二分类交叉熵，这个损失函数又称为重构损失（Reconstruction Loss）函数，公式如下：

$$Loss_{reconstruction} = \frac{1}{N}\sum_{i=1}^{N}\sum_{x=1}^{W}\sum_{y=1}^{H}\sum_{c=1}^{C} binary\_crossentropy(x_{x,y,c}, \hat{y}_{x,y,c})$$

x：输入图像。
$\hat{y}$：输出图像。
W：图像长度。
H：图像高度。
C：图像深度（RGB 或灰度）。
N：一个批量的数据量。

但是损失函数只有重构损失是不够的，因为图 11-4 介绍过 $\exp(\sigma^2)$ 用于控制 Noise（$\exp(\sigma^2) \times e$）的放大倍率，而 σ^2 又是从神经网络学习而来的，所以只要网络学会将 $\exp(\sigma^2)$ 输出为 0，就不会有 Noise（噪音）产生，如果没有噪音，就类似于自编码器架构。$\exp(\sigma^2)$ 的曲线图如图 11-8 所示，σ^2 越小，$\exp(\sigma^2)$ 就会越趋近于 0。

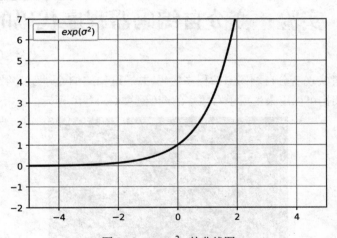

图 11-8 $\exp(\sigma^2)$ 的曲线图

为了解决这个问题，可以加上一个 $Loss_{\sigma 2}$ 来限制 σ^2，而如果要使 $Loss_{\sigma 2}=0$，σ^2 就必须为 0，此时 Noise 的放大倍率 $\exp(\sigma^2)=1$，因此解决了神经网络会往 $\exp(\sigma^2)=0$ 的方向更新的问题。损失函数曲线图如图 11-9 所示，公式如下：

$$Loss_{\sigma 2} = \frac{1}{2N}\sum_{i=1}^{N}[\exp(\sigma_i^2) - (1+\sigma_i^2)]$$

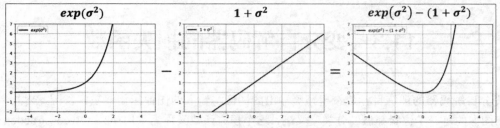

图 11-9　$exp(\sigma^2)-(1+\sigma^2)$ 的曲线图

另外，再对 μ 进行 L2 正则化（L2 Regularization），希望 μ 越小越好，公式如下：

$$\text{Loss}_\mu = \frac{1}{2N}\sum_{i=1}^{N}\mu_i^2$$

最后，将 Loss_μ 和 Loss_{σ^2} 损失函数整理成一个公式 $\text{Loss}_{\mu,\sigma^2}$（又称为 KL Loss，Kullback-Leibler Divergence Loss，即 KL 散度损失），公式如下：

$$\text{Loss}_{\mu\sigma^2} = \frac{1}{2N}\sum_{i=1}^{N}[\exp(\sigma_i^2)-(1+\sigma_i^2)+\mu_i^2]$$

μ：均值（编码器的其中一个输出）。
σ^2：方差（编码器的其中一个输出）。
N：一个批量的数据量。

11.4　实验：变分自编码器程序代码的实现

本节会带领读者实现变分自编码器的程序代码，并通过 MNIST 手写数字数据集训练网络模型，训练完成后的网络架构可以用来生成手写数字的图像，如图 11-10 所示。

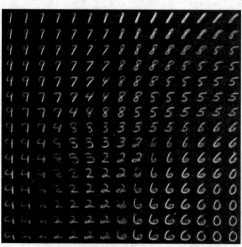

图 11-10　变分自编码器生成的手写数字图像

11.4.1 建立项目

由于本节的变分自编码器范例程序比前面第 1~9 章的范例程序复杂，因此使用 PyCharm IDE 来编写程序代码，建立项目的流程如下：

Step 01 使用 PyCharm 建立新项目。依次单击菜单选项 File→New Project，如图 11-11 所示。

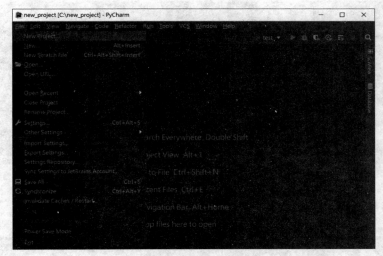

图 11-11　载入专案操作

Step 02 设置新项目的目录，如图 11-12 所示。

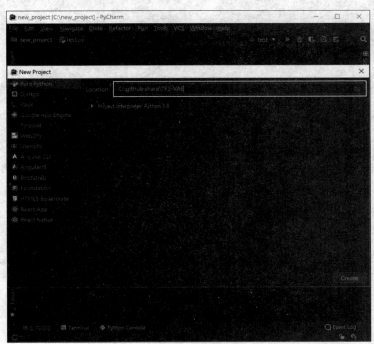

图 11-12　设置新项目的目录

Step 03 设置编译环境。首先打开 Project Interpreter：Python 3.6，选择 Existing interpreter 选项，并设置 Interpreter（Python 解释器），如图 11-13 所示。

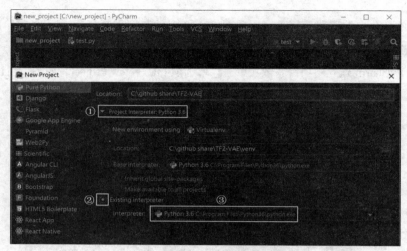

图 11-13　设置项目使用的 Interpreter

Step 04 建立项目，如图 11-14 所示。

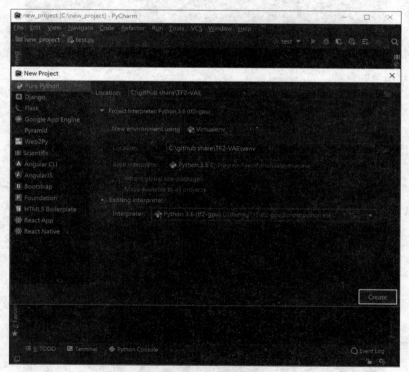

图 11-14　建立项目

说 明

本节的程序项目可以到 GitHub 网站的 https://github.com/KUASWoodyLIN/TF2-VAE 网页下载，如图 11-15 所示。

图 11-15　GitHub 网站上变分自编码器项目的页面

11.4.2　数据集介绍

本节的变分自编码器项目使用 MNIST 手写数字数据集，这个数据集拥有 60 000 训练数据和 10 000 测试数据，并且图像大小为 28×28×1，为灰度图像。可以通过 TensorFlow datasets 来下载并加载数据集，程序代码如下：

```
import tensorflow_datasets as tfds

# 加载训练数据集
train_data, info = tfds.load("mnist", split= tfds.Split.TRAIN, with_info=True)
# 加载测试数据集
test_data = tfds.load("mnist", split= tfds.Split.TRAIN)
# 显示数据集的信息
print(info)
```

结果如下：

```
tfds.core.DatasetInfo(
    name='mnist',
    version=1.0.0,
    description='The MNIST database of handwritten digits.',
    urls=['https://storage.googleapis.com/cvdf-datasets/mnist/'],
    features=FeaturesDict({
        'image': Image(shape=(28, 28, 1), dtype=tf.uint8),
        'label': ClassLabel(shape=(), dtype=tf.int64, num_classes=10),
```

```
        }),
        total_num_examples=70000,
        splits={
            'test': 10000,
            'train': 60000,
        },
        supervised_keys=('image', 'label'),
        citation="""@article{lecun2010mnist,
            title={MNIST handwritten digit database},
            author={LeCun, Yann and Cortes, Corinna and Burges, CJ},
            journal={ATT Labs [Online].Available: http://yann.lecun.com/exdb/mnist},
            volume={2},
            year={2010}
        }""",
        redistribution_info=,
)
```

11.4.3 变分自编码器项目说明

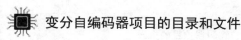 变分自编码器项目的目录和文件

建立变分自编码器项目的文件夹和 Python 文件，如图 11-16 所示。下面简单介绍需要建立的文件和目录。

图 11-16 变分自编码器项目的目录

- train.py

训练变分自编码器模型的程序代码。

- test.py

测试变分自编码器模型的程序代码。

- utils
  ➢ model.py：变分自编码器网络模型和自定义网络层的程序代码。

➢ losses.py：自定义损失函数的程序代码。
➢ callbacks.py：自定义回调函数的程序代码。

变分自编码器网络架构

首先在 utils→models.py Python 文件中编写变分自编码器的网络架构，这里建立一个 create_vae_model 的函数，其中包含编码器和解码器的网络模型，还有 $Loss_{u,\sigma^2}$ 的损失函数，网络架构如图 11-17 所示。

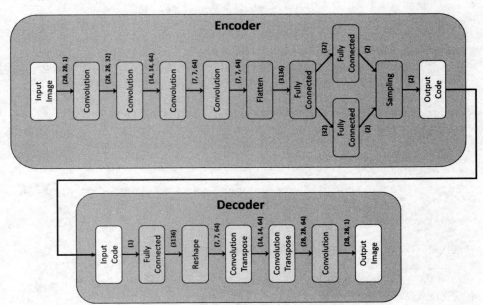

图 11-17　变分自编码器网络架构图

建立变分自编码器网络模型的函数，它的网络架构对照图 11-17：

```
def create_vae_model(input_shape, latent_dim):
    # 定义编码器网络模型
    img_inputs = keras.Input(input_shape)
    x = keras.layers.Conv2D(32, 3, padding='same', activation='relu')(img_inputs)
    x = keras.layers.Conv2D(64, 3, strides=2, padding='same', activation='relu')(x)
    x = keras.layers.Conv2D(64, 3, strides=2, padding='same', activation='relu')(x)
    x = keras.layers.Conv2D(64, 3, padding='same', activation='relu')(x)
    # 存储压平（Flatten）前的特征图大小，之后解码器还原时会用到
    shape_before_flatten = x.shape
    x = keras.layers.Flatten()(x)
    x = keras.layers.Dense(32, 'relu')(x)
    # 输出平均值 μ
    z_mean = keras.layers.Dense(latent_dim)(x)
```

```
    # 输出方差 σ*σ
    z_log_var = keras.layers.Dense(latent_dim)(x)
    # 自定义网络层,下面会说明
    z = Sampling()([z_mean, z_log_var])
    # 建立编码器网络模型
    encoder = keras.Model(inputs=img_inputs, outputs=z, name='encoder')
    encoder.summary()

    # 定义解码器网络模型
    latent_inputs = keras.Input((latent_dim,))
    # 产生和解码器压平前一样大小的特征
    x = keras.layers.Dense(np.prod(shape_before_flatten[1:]),
                        activation='relu')(latent_inputs)
    # 将特征重塑(Reshape)成编码器压平前的形状
    x = keras.layers.Reshape(target_shape=shape_before_flatten[1:])(x)
    x = keras.layers.Conv2DTranspose(64, 3, 2, padding='same',
activation='relu')(x)
    x = keras.layers.Conv2DTranspose(64, 3, 2, padding='same',
activation='relu')(x)
    img_outputs = keras.layers.Conv2D(1, 3, padding='same',
activation='sigmoid')(x)
    # 建立解码器网络模型
    decoder = keras.Model(inputs=latent_inputs, outputs=img_outputs,
                        name='decoder')
    decoder.summary()

    # 连接编码器和解码器,建立变分自编码器网络模型
    z = encoder(img_inputs)
    img_outputs = decoder(z)
    # 建立变分自编码器网络模型
    vae = keras.Model(inputs=img_inputs, outputs=img_outputs, name='vae')

    # Loss$_{u,\sigma^2} = \frac{1}{2N}\sum_{i=1}^{N}[\exp(\sigma_i^2)-(1+\sigma_i^2)+\mu_i^2]$,又称为 KL Loss,对变分自编码器中间层的输出计算损失值。
    # 这里用 vae.add_loss 来加入损失函数
    kl_loss = 0.5 * tf.reduce_mean(tf.exp(z_log_var) - (1 + z_log_var) +
                        tf.square(z_mean))
    vae.add_loss(kl_loss)
    return vae
```

其中,Sampling 为自定义网络层,如图 11-18 所示,矩形框内的运算为 Sampling 层。Sampling 自定义网络层的程序代码如下:

```
class Sampling(keras.layers.Layer):
    def call(self, inputs):
        z_mean, z_log_var = inputs
        # 批量的大小
        batch = tf.shape(z_mean)[0]
        # 编码的维度
        dim = tf.shape(z_mean)[1]
        # 产生标准正态分布
```

```
epsilon = tf.random.normal(shape=(batch, dim))
return z_mean + tf.exp(z_log_var) * epsilon
```

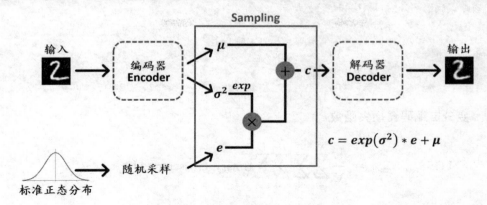

图 11-18　变分自编码器示意图

> **说　明**
>
> $Loss_{u,\sigma^2}$（KL Loss）损失函数是对变分自编码器网络中间输出的 μ 和 σ 计算损失，而这个运算发生在变分自编码器网络中间，所以需要在建立网络的层一同声明内部的损失值。上面介绍的 vae.add_loss 为其中一种方法，我们也可以建立自定义网络模型或自定义网络层来加入 $Loss_{u,\sigma^2}$ 损失函数。

范例 1：自定义网络模型

```python
class VariationalAutoEncoder(keras.Model):
    def __init__(self, name='autoencoder', **kwargs):
        super(VariationalAutoEncoder, self).__init__(name=name, **kwargs)
        self.encoder = Encoder()
        self.decoder = Decoder()
        self.sampling = Sampling()

    def call(self, inputs):
        z_mean, z_var = self.encoder(inputs)
        z = self.sampling([z_mean, z_var])
        img_output = self.decoder(z)
        kl_loss=0.5*tf.reduce_mean(tf.exp(z_var)-(1+z_var)+
                            tf.square(z_mean))
        self.add_loss(kl_loss) return reconstructed
```

范例 2：自定义网络层

```python
class KLLoss(keras.layers.Layer):
    def call(self, inputs):
        z_mean = inputs[0]
        z_var = inputs[1]
        kl_loss=0.5*tf.reduce_mean(tf.exp(z_var)-(1+z_var)+
                            tf.square(z_mean))
        self.add_loss(kl_loss)
```

```
            return z_mean, z_var

# … 省略编码器部分的卷积层和全连接层 …
z_mean = keras.layers.Dense(latent_dim)(x)
z_var = keras.layers.Dense(latent_dim)(x)
z_mean, z_var = KLLoss()([z_mean, z_var])
z = Sampling()([z_mean, z_var])
encoder = keras.Model(inputs=img_inputs, outputs=z, name='encoder')
```

变分自编码器损失函数

$$\text{LOSS}_{\text{reconstruction}} = \frac{1}{N} \sum_{i=1}^{N} \sum_{x=1}^{W} \sum_{y=1}^{H} \sum_{c=1}^{C} \text{Binary\_crossentropy}(x_{x,y,c}, \hat{y}_{x,y,c})$$

x：输入图像。
$\hat{y}$：输出图像。
W：图像长度。
H：图像高度。
C：图像深度（RGB 或灰度）。
N：一个批量的数据量。

要定制重构损失函数，可以把程序代码编写在 utils→losses.py 的 Python 文件中：

```python
def reconstruction_loss(y_true, y_pred):
    # 对生成图像与输入图像的每一个像素计算二分类交叉熵
    bce = -(y_true * tf.math.log(y_pred + 1e-07) +
        (1 - y_true) * tf.math.log(1 - y_pred + 1e-07))
    # tf.reduce_sum:将每一个像素加总, tf.reduce_mean:计算批量数据的平均值
    return tf.reduce_mean(tf.reduce_sum(bce, axis=[1, 2, 3]))
```

自定义回调函数

这部分程序会建立两个自定义回调函数，分别如下，可以把程序代码编写在 utils→callbacks.py 的 Python 文件中。

- SaveDecoderModel

每一个 epoch 都会检查网络模型有没有进步，如果有，就存储变分自编码器中的解码器模型（类似于 keras.callbacks.ModelCheckpoint）。

- SaveDecoderOutput

每一个 epoch 都会生成 225 张图像，并存储在 TensorBoard 记录文件中，之后可以启动 TensorBoard 观察每一个 epoch 的输出变化。

自定义回调函数 SaveDecoderModel 代码如下：

```python
class SaveDecoderModel(tf.keras.callbacks.Callback):
```

```python
    def __init__(self, weights_file, monitor='loss', save_weights_only
=False):
        super(SaveDecoderModel, self).__init__()
        self.weights_file = weights_file          # 解码器网络模型存储的路径
        self.best = np.Inf                        # 设置 best 为无限大
        self.monitor = monitor    # 要监测的名称(ex: loss or val_loss)
        self.save_weights_only = save_weights_only  # 存储模型或模型权重

    def on_epoch_end(self, epoch, logs=None):
        """
        每一个 epoch 会执行一次，如果网络有进步，就将网络模型或权重存储起来
        """
        loss = logs.get(self.monitor)             # 取得要监测的数值
        if loss < self.best:
            if self.save_weights_only:
                # 存储解码器网络模型的权重
                self.model.get_layer('decoder').save_weights
(self.weights_file)
            else:
                # 存储完整的解码器网络模型
                self.model.get_layer('decoder').save(self.weights_file)
            self.best = loss
```

自定义回调函数 SaveDecoderOutput 代码如下：

```python
class SaveDecoderOutput(tf.keras.callbacks.Callback):
    def __init__(self, image_size, log_dir):
        super(SaveDecoderOutput, self).__init__()
        self.size = image_size    # 生成图像的大小
        self.log_dir = log_dir    # Tensorboard 记录文件的存储路径
        n = 15                    # 生成(15×15)张图像
        # 图像存储及显示的大数组（可以放入 225 张图像）
        self.save_images = np.zeros((image_size * n, image_size * n, 1))
        # 线性抽取 15 个数值，作为编码的 x
        self.grid_x = np.linspace(-1.5, 1.5, n)
        # 线性抽取 15 个数值，作为编码的 y
        self.grid_y = np.linspace(-1.5, 1.5, n)

    def on_train_begin(self, logs=None):
        """训练开始前会建立 Tensorboard 记录文件"""
        path = os.path.join(self.log_dir, 'images')
        self.writer = tf.summary.create_file_writer(path)

    def on_epoch_end(self, epoch, logs=None):
        """
        每一个 epoch 会执行一次，每次生成 225 张图像，并存储在记录文件中
        """
        for i, yi in enumerate(self.grid_x):
            for j, xi in enumerate(self.grid_y):
                # 产生一组编码
                z_sample = np.array([[xi, yi]])
```

```
            # 解码器通过编码生成图像
            img = self.model.get_layer('decoder')(z_sample)
            # 将图像存储到图像存储显示的数组中
            self.save_images[i*self.size:(i+1)*self.size,
                             j*self.size:(j+1)*self.size] = img.numpy()[0]
    # 将生成的 225 张图像存储到 TensorBoard 记录文件中
    with self.writer.as_default():
        tf.summary.image("Decoder output", [self.save_images], step=epoch)
```

11.4.4 变分自编码器训练和生成图像

训练变分自编码器

这部分程序会训练变分自编码器的网络模型，可以把程序代码编写在 train.py 文件中。

Step 01 导入必要套件。

```
import os
import tensorflow as tf
import tensorflow_datasets as tfds
from tensorflow import keras
from utils.models import create_vae_model
from utils.losses import reconstruction_loss
from utils.callbacks import SaveDecoderOutput, SaveDecoderModel
```

Step 02 数据预处理。

将图像标准化到 0~1。此外，由于变分自编码器的训练方式是希望输出的预测与输入相同，因此训练数据（x）和训练答案（y）是相同的图像数据。

```
def parse_fn(dataset, input_size=(28, 28)):
    x = tf.cast(dataset['image'], tf.float32)
    # 将图像大小重新调整成神经网络输入的大小
    x = tf.image.resize(x, input_size)
    # 将图像进行标准化，缩放到 0~1
    x = x / 255.
    # 返回训练数据和训练答案
    return x, x
```

Step 03 加载 MNIST 数据集。

```
train_data = tfds.load('mnist', split=tfds.Split.TRAIN)
test_data = tfds.load('mnist', split=tfds.Split.TEST)
```

Step 04 设置数据集。

```
AUTOTUNE = tf.data.experimental.AUTOTUNE              # 自动调整模式
batch_size = 16   # 批量大小
train_num = info.splits['train'].num_examples         # 训练数据的数量
```

```
# 打乱数据集
train_data = train_data.shuffle(train_num)
# 加载预处理 parse_fn 函数,CPU 数量为自动调整模式
train_data = train_data.map(parse_fn, num_parallel_calls=AUTOTUNE)
# 设置批量大小为 16,并开启预取模式(缓存空间为自动调整模式)
train_data = train_data.batch(batch_size).prefetch(buffer_size=AUTOTUNE)

# 加载预处理 parse_fn 函数,CPU 数量为自动调整模式
test_data = test_data.map(parse_fn, num_parallel_calls=AUTOTUNE)
# 设置批量大小为 16,并开启预取模式(缓存空间为自动调整模式)
test_data = test_data.batch(batch_size).prefetch(buffer_size=AUTOTUNE)
```

Step 05 建立回调函数。

```
# 创建存储模型权重的目录
log_dirs = 'logs_vae'
model_dir = log_dirs + '/models'
os.makedirs(model_dir, exist_ok=True)

# 将训练记录存成 TensorBoard 的记录文件
model_tb = keras.callbacks.TensorBoard(log_dir=log_dirs)
# 存储最好的网络模型权重
model_sdw = SaveDecoderModel(model_dir + '/best_model.h5',
                             monitor='val_loss')
# 把解码器生成的图像存储到 TensorBoard 记录文件中
model_testd = SaveDecoderOutput(28, log_dir=log_dirs)
```

Step 06 建立变分自编码器网络模型。

```
# 定义网络输入图像的大小
input_shape = (28, 28, 1)
# 定义编码器压缩到多少维的空间向量
latent_dim = 2
vae_model = create_vae_model(input_shape, latent_dim)
```

Step 07 设置训练使用的优化器和损失函数。

```
optimizer = tf.keras.optimizers.RMSprop()
vae_model.compile(optimizer, loss=reconstruction_loss)
```

Step 08 训练网络模型。

```
vae_model.fit(train_data,
              epochs=20,
              validation_data=test_data,
              callbacks=[model_tb, model_sdw, model_testd])
```

解码器生成图像

这部分程序会测试刚刚训练的变分自编码器网络模型的解码器网络,可以把程序代码编写在 test.py 文件中,测试结果如图 11-19 所示。

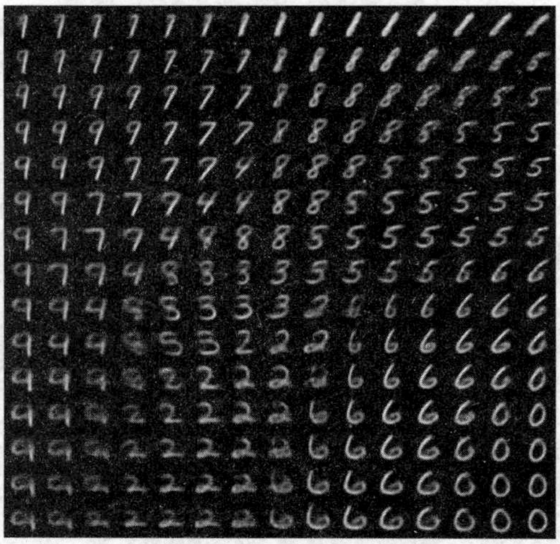

图 11-19 解码器的预测结果，总共预测出 225 张图像

Step 01 导入必要套件。

```
import numpy as np
import tensorflow as tf
import matplotlib.pyplot as plt
```

Step 02 解码器网络模型生成图像。

```
size = 28       # 输出图像的大小
n = 15          # 生成(15×15)张图像
# 图像存储及显示的大数组(可以放入 225 张图像)
save_images = np.zeros((size * n, size * n, 1))
grid_x = np.linspace(-1.5, 1.5, n) # 线性抽取 15 个数值作为编码的 x
grid_y = np.linspace(-1.5, 1.5, n) # 线性抽取 15 个数值作为编码的 y

# 加载刚刚训练好的解码器网络模型
model = tf.keras.models.load_model('logs_vae/models/best_model.h5')
for i, yi in enumerate(grid_x):
    for j, xi in enumerate(grid_y):
        # 生成一组编码
        z_sample = np.array([[xi, yi]])
        # 解码器通过编码生成图像
        img = model(z_sample)
        # 将图像存储到图像存储显示的数组中
        save_images[i * size: (i + 1) * size, j * size: (j + 1) * size] = img.numpy()[0]

# 显示解码器的预测结果，总共预测出 225 张图片
plt.imshow(save_images[..., 0], cmap='gray')
plt.show()
```

用 TensorBoard 观察训练

最后可以通过 TensorBoard 来观察解码器训练的预测变化，如图 11-20 所示，这个图像记录是通过前面的 SaveDecoderOutput 自定义回调函数实现的功能。

启动 TensorBoard（命令行）：

```
tensorboard --logdir logs-vae
```

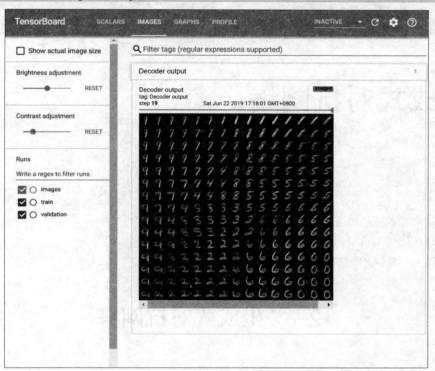

图 11-20　用 TensorBoard 观察解码器训练的预测变化

11.5　参考文献

[1] Hinton G, Salakhutdinov R. Reducing the dimensionality of data with neural networks [J]. Science, 2006, 313(5786): 504-507.

[2] Hinton G E, Zemel R. Autoencoders, minimum description length and helmholtz free energy [C]. Advances in Neural Information Processing Systems, 1993, 3-10.

[3] Vincent P, Larochelle H, Lajoie I, et al. Stacked denoising autoencoders: Learning useful representations in a deep network with a local denoising criterion [J]. The Journal of Machine Learning Research, 2010.

[4] Kingma D P, Welling M. Auto-encoding variational bayes [C]. In International Conference on Learning Representations, 2014.

第 12 章

生成式对抗网络

学习目标

- 介绍生成式对抗网络的网络架构
- 介绍改进的 WGAN（WGAN-GP）网络架构
- 实现 WGAN-GP 项目来生成人脸图片

12.1 生成式对抗网络

12.1.1 生成式对抗网络介绍

生成式对抗网络（Generative Adversarial Network，GAN）[1]是 2014 年由 Ian Goodfellow 提出来的方法，GAN 的架构是由生成器（Generator）和判别器（Discriminator）所组成的。如图 12-1 所示，在 GAN 的架构中，生成器负责生成图像，判别器负责判别图像的真实性，真实图像给予高分，生成图像给予低分。生成图像越接近真实图像，判别器就会给生成图像越高的分数，判别器的输出会经过 Sigmoid 激活函数，所以分数会介于 0~1。

生成式对抗网络的工作过程可以想象成，一位学生（Generator）和一位老师（Discriminator），学生会不断地画人像，而老师会对学生的作品和真实的人像进行判别，也就是判别图像的"真假程度"，然后学生就会根据老师的判别结果不断改进作品，而老师也会根据学生作品状况来增加判别的难度，形成互相对抗成长的循环。训练到最后，学生的作品与真实的人像难以分辨，这就是生成式对抗网络的主要概念，如图 12-2 所示。

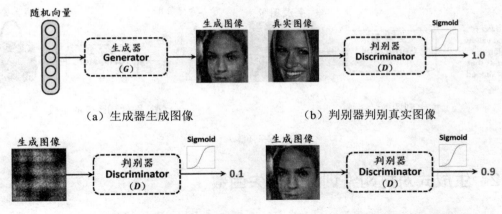

图 12-1　生成器和判别器的工作流程示意图

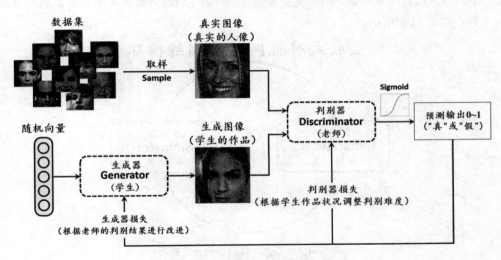

图 12-2　生成式对抗网络的架构

以生成式对抗网络为概念，近几年又延伸出了 DCGAN（2015）[2]、Improved GAN（2016）[3]、PACGAN（2017）[4]、WGAN（2017）[5]、WGAN-GP（2017）[6]、CycleGAN（2017）[7]、PGGAN（2017）[8]、StackGAN（2017）[9]、video2video（2018）[10]、BigGAN（2019）[11]、StyleGAN（2019）[12]，而且衍生出了非常多的应用，例如图 12-3（a）所示的生成图像，图 12-3（b）所示的生成文字，图 12-3（c）所示的文字到图像的生成，图 12-3（d）所示的图像到图像的生成，等等。

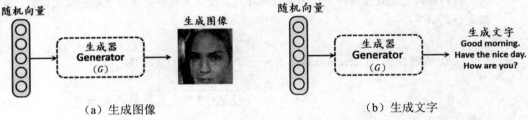

图 12-3　生成式对抗网络各种应用

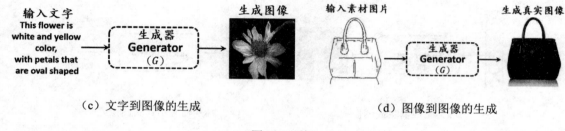

（c）文字到图像的生成　　　　　　（d）图像到图像的生成

图 12-3（续）

12.1.2　生成式对抗网络训练及损失函数

生成式对抗网络的训练方式与前几章介绍的网络架构训练方法不同，必须将生成器和判别器分开训练。在训练生成器时，使用生成器损失函数进行训练，而训练判别器时，则使用判别器损失函数进行训练。生成器和判别器之间是交互训练的，例如先训练判别器 N 次，再训练生成器一次，N 为超参数，如图 12-4 所示。

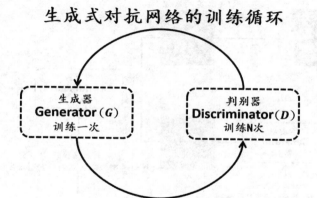

图 12-4　生成式对抗网络的训练循环

下面介绍生成器训练和判别器训练的损失函数。

- 生成器训练的目的是追求最低的生成器损失（Generator Loss）。训练时需将判别器的权重固定住，只训练生成器的权重，生成器损失越低，代表判别器认为生成器生成的图像越接近真实。如图 12-5 所示，一组随机向量（z）进入生成器后，得到生成图像（$\hat{x}$），生成图像再通过判别器预测后，预测输出越接近 1 越好，而判别器预测输出越接近 1，代表生成器损失越低，判别器认为生成器网络产生的图像越真实，生成器越分辨不出生成图像的真假。生成器损失公式如下：

$$\text{生成器损失（Generator Loss）} = \frac{1}{N}\sum_{i=1}^{N}\log(1-\boxed{D(G(Z^i))})$$

越接近1则生成器损失会越低

N：训练数据量。

D：判别器。
G：生成器。
z：从一个分布中抽样作为生成器的输入。

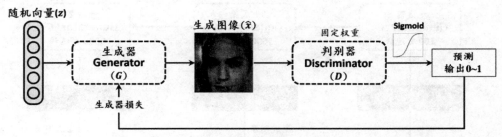

图 12-5　生成器训练示意图

- 判别器训练的目的是追求最低的判别器损失（Discriminator Loss）。训练判别器时，需将生成器的权重固定住，只训练判别器的权重，判别器损失越低，代表判别器越有能力分辨出真实图像与生成图像。如图 12-6 所示，真实图像（x）进入判别器后，预测输出越接近 1 越好；一组随机向量（z）进入生成器后，得到生成图像（$\hat{x}$），生成图像再通过判别器预测后，预测输出越接近 0 越好。判别器预测真实图像越接近 1、生成图像越接近 0，代表判别器损失会越低，判别器能够轻易分辨出真实图像与生成图像。判别器损失公式如下：

$$判别器损失（\text{Discriminator Loss}）= -\frac{1}{N}\sum_{i=1}^{N}\log \underbrace{D(x^i)}_{越接近1则判别器损失会越低} - \frac{1}{N}\sum_{i=1}^{N}\log(1-\underbrace{D(\hat{x}^i)}_{越接近0则判别器损失会越低})$$

N：训练数据量。
D：判别器。
x：从数据集中抽样的真实图像。
$\hat{x}$：从生成器产生的生成图像。

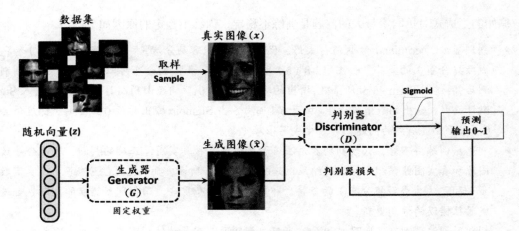

图 12-6　判别器训练示意图

对生成式对抗网络而言，一个迭代（Iterative）的训练，代表生成器和判别器至少都有被训练一次，例如一个迭代训练生成器 1 次、判别器 N 次。而每一次迭代后，生成器会产生更真实的生成图像，判别器也会将评判标准提高，如图 12-7 所示。

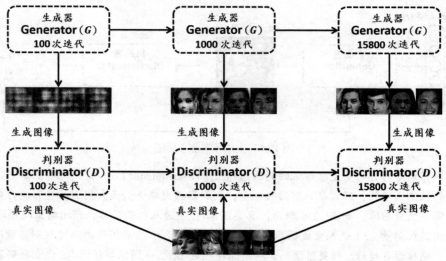

图 12-7　生成式对抗网络的训练循环

12.2　GAN、WGAN、WGAN-GP 的演进

本节主要介绍 Wasserstein GAN Improved[6]（又称作 WGAN-GP），在介绍 WGAN-GP 之前，我们先分析原始 GAN[1]和 WGAN[5]的问题，并介绍 WGAN-GP[6]的损失函数和改进原理。

12.2.1　生成式对抗网络的问题

原始的生成式对抗网络最大的问题是训练不稳定，训练不稳定的原因如下：

- 判别器（Discriminator）训练得太好，代表判别器很容易分辨 $\mathbb{P}_{data}$（真实数据分布）和 $\mathbb{P}_G$（生成数据分布）的差异，如图 12-8（a）所示，即 $\mathbb{P}_{data}$ 和 $\mathbb{P}_G$ 的分布可以分开，$\mathbb{P}_G$ 经由判别器预测后，再进入 Sigmoid 激活函数，输出的数值会接近 0，$\mathbb{P}_{data}$ 经由判别器预测后，再进入 Sigmoid 激活函数，输出的数值接近 1。由表 6-1 可知，当 Sigmoid 输出接近 0 或 1 时，梯度值会趋近 0，由于梯度消失的问题，导致生成器很难更新。

- 判别器训练得不好，代表判别器无法给予生成图像和真实图像正确的判断，可能会导致生成图像和真实图像预测出相似的结果，如图 12-8（b）所示，判别器无法分辨 $\mathbb{P}_{data}$（真实数据分布）和 $\mathbb{P}_G$（生成数据分布）的差异，即 $\mathbb{P}_{data}$ 和 $\mathbb{P}_G$ 分布重叠，而判别器错误的判断也会造成生成器往错误的方向更新。

- 判别器训练得刚好，如图 12-8（c）所示，判别器不容易训练到刚好的程度（避免发生图 12-8

（a）判别器训练得太好或图 12-8（b）判别器训练得不好的情况），因此在判别器训练时，需仔细观察判别器损失，这样才有机会让生成器生成更真实的图像。

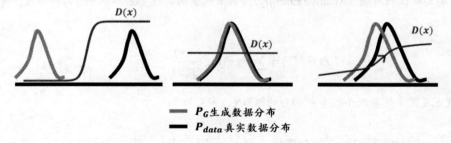

（a）判别器训练得太好　（b）判别器训练得不好　（c）判别器训练得刚好

图 12-8　训练不稳定的原因

GAN[1]论文提出了判别器的最大化目标函数，定义如下：

$$\max_D V(G,D) = \mathbb{E}_{x \sim \mathbb{P}_{data}}[\log(D(x))] + \mathbb{E}_{x \sim \mathbb{P}_G}[1 - D(x)] \tag{1}$$

在固定生成器的情况下，可以得到最佳的判别器如下：

$$D^*(x) = \frac{\mathbb{P}_{data}(x)}{\mathbb{P}_{data}(x) + \mathbb{P}_G(x)} \tag{2}$$

因此，公式（1）可以重写如下：

$$\begin{aligned}\max_D V(G,D) &= V(G,D^*) \\ &= \mathbb{E}_{x \sim \mathbb{P}_{data}}[\log \frac{\mathbb{P}_{data}(x)}{\mathbb{P}_{data}(x) + \mathbb{P}_G(x)}] + \mathbb{E}_{x \sim \mathbb{P}_G}[\log \frac{\mathbb{P}_{data}(x)}{\mathbb{P}_{data}(x) + \mathbb{P}_G(x)}] \\ &= -2\log 2 - KL(\mathbb{P}_{data} \| \frac{\mathbb{P}_{data} + \mathbb{P}_G}{2}) + KL(\mathbb{P}_G \| \frac{\mathbb{P}_{data} + \mathbb{P}_G}{2}) \\ &= -2\log 2 + 2JS(\mathbb{P}_{data} \| \mathbb{P}_G)\end{aligned} \tag{3}$$

$\mathbb{P}_{data}$：真实数据分布。

$\mathbb{P}_G$：生成数据分布。

D：判别器。

D^*：最佳化的判别器。

KL：Kullback-Leibler Divergence（KL 散度）。

JS：Jensen-Shannon Divergence（JS 散度）。

> **说　明**
>
> KL 散度又称作相对熵（Relative Entropy），用来衡量两个概率分布 $\mathbb{P}$ 和 $\mathbb{Q}$ 的相似度，定义如下：
>
> $$KL(\mathbb{P}\|\mathbb{Q}) = \sum_{x\in X}\mathbb{P}(x)\log\frac{\mathbb{P}(x)}{\mathbb{Q}(x)} \tag{4}$$
>
> JS 散度也是用来衡量两个概率 $\mathbb{P}$ 和 $\mathbb{Q}$ 的相似度，定义如下：
>
> $$JS(\mathbb{P}\|\mathbb{Q}) = \frac{1}{2}KL(\mathbb{P}\|\frac{\mathbb{P}+\mathbb{Q}}{2}) + \frac{1}{2}KL(\mathbb{Q}\|\frac{\mathbb{P}+\mathbb{Q}}{2}) \tag{5}$$

在公式（3）中，最佳的判别器是使用 JS 散度监测 $\mathbb{P}_{data}$（真实数据分布）和 $\mathbb{P}_G$（生成数据分布）的相似度。在图 12-9（a）中，当 $\mathbb{P}_{data}$ 和 $\mathbb{P}_G$ 两个分布一致时，JS 散度计算出的数值为 0，但在图 12-9（b）和图 12-9（c）中，当 $\mathbb{P}_{data}$ 和 $\mathbb{P}_G$ 两个分布不相交时，JS 散度计算出的数值都是固定常数 log2，因此 $\mathbb{P}_{data}$ 和 $\mathbb{P}_G$ 分布在不相交的情况下，生成器无法有效更新。

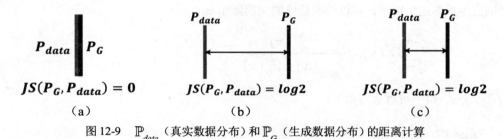

图 12-9　$\mathbb{P}_{data}$（真实数据分布）和 $\mathbb{P}_G$（生成数据分布）的距离计算

12.2.2　Wasserstein 距离介绍

为了解决 JS 散度监测两个分布之间的相似度问题，WGAN[5]论文建议移除判别器的 Sigmoid 激活函数，以避免梯度消失问题，并提出 Wasserstein 距离监测方法，用来监测两个概率分布之间的距离，定义如下：

$$W(\mathbb{P}_{data},\mathbb{P}_G) = \inf_{\gamma\sim\Pi(\mathbb{P}_{data},\mathbb{P}_G)}\mathbb{E}_{(x,y)\sim\gamma}[\|x-y\|] \tag{6}$$

$\mathbb{P}_{data}$：真实数据分布。
$\mathbb{P}_G$：生成数据分布。
$\Pi(\mathbb{P}_{data},\mathbb{P}_G)$：$\mathbb{P}_{data}$ 和 $\mathbb{P}_G$ 分布的所有可能的联合分布的集合。
x, y：从可能的联合分布 γ 采样得到的样本。

公式(6)可以比喻为推土机的铲土任务，Π 包含土堆所有可能的运输路径，$\mathbb{E}_{(x,y)\sim\gamma}[\|x-y\|]$ 代表在 γ 这个运输路径下，将土堆 $\mathbb{P}_{data}$ 搬移到土堆 $\mathbb{P}_G$ 所需要的距离消耗，而 Wasserstein 距离就是在最佳运输路径规划（Optimal Transport Plan）下的距离消耗。

Wasserstein 距离可根据 Kantorovich-Rubinstein Duality[13]定理,将公式(6)转换成下面的形式:

$$W(\mathbb{P}_{data},\mathbb{P}_G) = \max_{D\in 1-Lipschitz}\{\mathbb{E}_{x\sim \mathbb{P}_{data}}[D(x)] - \mathbb{E}_{x\sim \mathbb{P}_G}[D(x)]\} \tag{7}$$

$\mathbb{P}_{data}$:真实数据分布。
$\mathbb{P}_G$:生成数据分布。
D:判别器。
1-Lipschitz:主要是限制判别器,$\|D(x_1)-D(x_2)\| \leq k\|x_1-x_2\|$,k=1。判别器为一个复杂函数,并且满足 1-Lipschitz,即判别器函数的最大斜率不能超过 1。

> **说 明**
>
> 从直观角度说明判别器为什么需要满足 1-Lipschitz 函数,如果判别器不满足 1-Lipschitz 函数,判别器就会将$\mathbb{P}_G$往-∞更新、将$\mathbb{P}_{data}$往+∞更新,训练到最后就会发生崩溃,如图 12-10 所示。 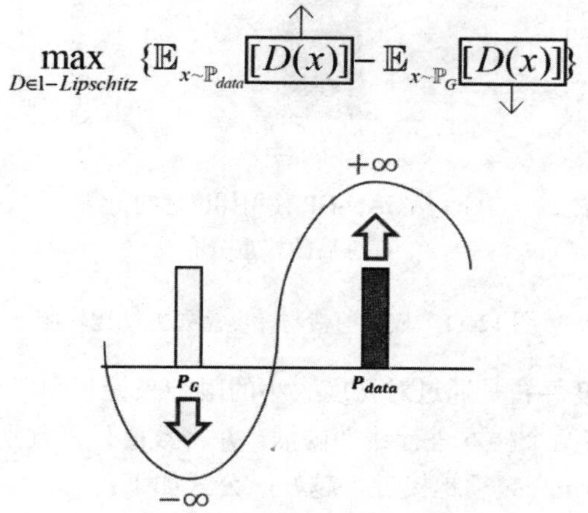
>
> 图 12-10 判别器的输出情况

WGAN 使用权重剪裁(Weight Clipping)的方式限制判别器的权重大小,让更新后的判别器权重必须在超参数 c 的控制范围内(如果权重大于 c,就设置为 c,如果权重小于-c,就设置为-c,否则权重保持原数值),通过权重剪裁的方式让判别器满足 1-Lipschitz 函数,公式如下:

$$w \leftarrow clip(w,-c,c) \tag{8}$$

w:判别器权重。
c:限制权重极值的超参数。

但是,WGAN 使用的权重剪裁方法强制满足判别器为 1-Lipschitz 函数,会遇到几个问题,问题如下:

- 如图 12-11(a)所示,WGAN 将权重剪裁方法用于 Swiss Roll 数据集的训练上,并限制判别器权重的极大值和极小值为-0.01和0.01，而经过训练后发现，判别器的绝大多数权重会集中在极大值和极小值，这样导致模型无法描述复杂的问题。

- 如图12-11（c）所示，WGAN将权重剪裁方法用于Swiss Roll数据集的训练上，如果超参数c设置得稍微大一点，每经过一层网络层，梯度就会变大一点，经过多层传递后就会发生梯度爆炸问题；反之，如果超参数c设置得稍微小一点，每经过一层网络层，梯度就会变小一点，经过多层传递后，就会发生梯度消失问题。

因此，WGAN-GP提出用梯度惩罚（Gradient Penalty）的方法来解决上述问题，并且满足1-Lipschitz。如图12-11（b）所示，使用梯度惩罚方法训练判别器后的权重分布更具有多样性，可以让模型描述更复杂的问题；如图12-11（c）所示，使用梯度惩罚方法可以有效改善梯度消失与梯度爆炸的问题。

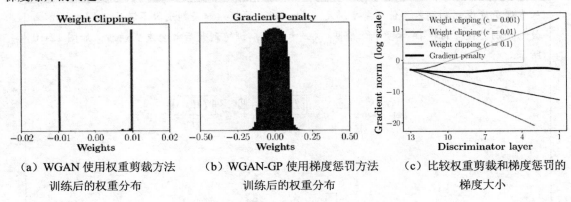

（a）WGAN 使用权重剪裁方法训练后的权重分布　　（b）WGAN-GP 使用梯度惩罚方法训练后的权重分布　　（c）比较权重剪裁和梯度惩罚的梯度大小

图 12-11　权重剪裁与梯度惩罚方法比较图

WGAN-GP 使用的梯度惩罚是通过对梯度大小的限制来让判别器满足 1-Lipschitz 函数，它的做法是在 $\mathbb{P}_{data}$ 和 $\mathbb{P}_G$ 抽样后将两个样本进行线性内插，并将数值代入判别器计算梯度，而梯度值的欧几里德范数（Euclidean Norm）必须越接近1越好，公式如下：

$$\max_D \{\mathbb{E}_{x \sim \mathbb{P}_{data}}[D(x)] - \mathbb{E}_{x \sim \mathbb{P}_G}[D(x)] - \underbrace{\lambda \mathbb{E}_{x \sim \mathbb{P}_{penalty}}[(\|\nabla_x D(x)\|_2 - 1)^2]}_{\text{梯度惩罚}}\} \tag{9}$$

$\mathbb{P}_{data}$：真实数据分布。
$\mathbb{P}_G$：生成数据分布。
$\mathbb{P}_{penalty}$：从真实数据分布（$\mathbb{P}_{data}$）和生成数据分布（$\mathbb{P}_G$）中抽样，将两个样本进行线性内插。
D：判别器。

12.2.3　WGAN-GP 损失函数

WGAN-GP 与原始生成式对抗网络的训练流程相同，它将原 GAN 架构的判别器最后一层的 Sigmoid 激活函数移除了，并且换成线性输出，并对判别器损失和生成器损失都进行了改进，损失函数分别如下：

- 生成器训练的目的是追求最低的生成器损失，训练时需将判别器的权重固定住，只训练生成器的权重，生成器损失越低，代表判别器认为生成器生成的图像越接近真实。例如，一组随机向量（z）进入生成器后，得出生成图像（$\hat{x}$），生成图像再通过判别器预测后，预测输出越大越好，而判别器预测输出越大，代表生成器损失越低，判别器认为生成器网络生成的图像越真实，判别器越分辨不出生成图像的真假。生成器损失公式如下：

$$\text{生成器操损失（Generator Loss）} = -\frac{1}{N}\sum_{i=1}^{N} \underbrace{D(G(z^i))}_{\text{越大则损失越低}}$$

N：训练数据量。
D：判别器。
G：生成器。
z：从一个分布中抽样作为生成器的输入。

- 判别器训练的目的是追求最低的判别器损失。训练判别器时，需将生成器的权重固定住，只训练判别器的权重，判别器损失越低，代表判别器越有能力分辨出真实图像与生成图像。例如，真实图像（x）进入判别器后，预测输出越大越好；一组随机向量（z）进入生成器后得出生成图像（$\hat{x}$），生成图像再通过判别器预测后，预测输出越小越好。判别器预测真实图像越大，预测生成图像越小，代表判别器损失会越低，判别器能够轻易分辨真实图像与生成图像。判别器损失公式如下：

$$\text{判别器操损失（Discriminator Loss）} = -\frac{1}{N}\sum_{i=1}^{N}\underbrace{D(x^i)}_{\text{越大则损失值越低}} + \frac{1}{N}\sum_{i=1}^{N}\underbrace{D(\hat{x}^i)}_{\text{越小则损失值越低}} + \lambda \times Gradient\ Penalty$$

N：训练数据量。
D：判别器。
x：从数据集中抽样的真实图像。
$\hat{x}$：从生成器得出的生成图像。
λ：梯度惩罚系数，通常设置为10。

- 梯度惩罚方法通过对梯度的惩罚来约束判别器，并满足 1-Lipschitz 函数。在 $\mathbb{P}_{data}$（真实数据分布）和 $\mathbb{P}_G$（生成数据分布）抽样后，将两个样本进行线性内插，并将数值代入判别器计算梯度，而梯度的欧几里得范数必须越接近 1 越好，公式如下：

$$Gradient\ Penalty = \frac{1}{N}\sum_{i=1}^{N}\underbrace{(\|\nabla_{\tilde{x}}D(\tilde{x})\|_2 - 1)^2}_{\text{越接近1损失值越低}}$$

$$\tilde{x} = tx + (1-t)\hat{x}$$

N:训练数据量。
D:判别器。
$\hat{x}$:从真实图像(x)和生成图像($\tilde{x}$)中抽样,将两个样本进行线性内插。
t:为 0~1 均匀分布的随机抽样数值。

12.3 实验:WGAN-GP 程序代码的实现

在介绍 WGAN-GP 后,本节主要实现 WGAN-GP 的程序代码,并通过 Large-Scale CelebFaces Attributes Dataset(CelebA,一个大型人脸属性数据集)训练网络模型,最后训练完成的网络架构可以用来生产人脸图像。图 12-12(a)所示为训练 WGAN-GP 网络 100 次迭代后输出的图像,图 12-12(b)所示为训练 WGAN-GP 网络 1000 次迭代后输出的图像,图 12-12(c)所示为训练 WGAN-GP 网络 15800 次迭代后输出的图像。

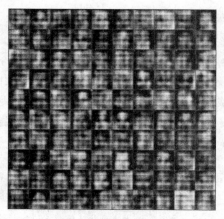

(a)训练 WGAN-GP 网络 100 次迭代后输出的图像

(b)训练 WGAN-GP 网络 1000 次迭代后输出的图像　　(c)训练 WGAN-GP 网络 15800 次迭代输出的图像

图 12-12　训练 WGAN-GP 网络后输出的人脸图像

12.3.1 建立项目

由于本节的 WGAN-GP 范例程序比前面第 1~9 章的范例程序都要复杂，因此我们使用 PyCharm IDE 来编写程序代码，建立项目的流程如下：

Step 01 使用 PyCharm 建立新项目。依次单击菜单选项 File→New Project，如图 12-13 所示。

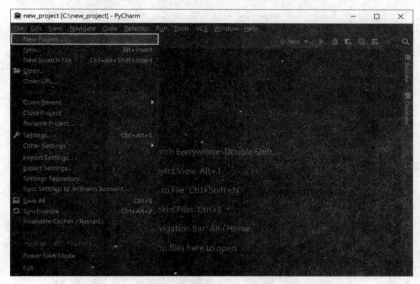

图 12-13　WGAN-GP GitHub 项目页面

Step 02 设置新项目的目录，如图 12-14 所示。

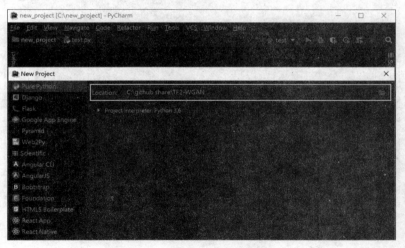

图 12-14　设置新项目的目录

Step 03 设置编译环境。首先打开 Project Interpreter：Python 3.6，选择 Existing interpreter 选项，并设置 Interpreter（Python 解释器），如图 12-15 所示。

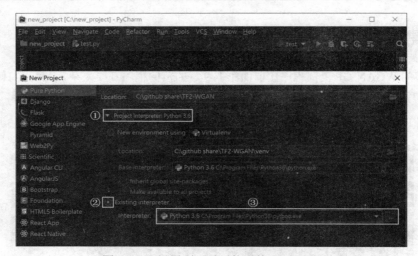

图 12-15　设置项目需要使用的 Interpreter

Step 04 建立项目，如图 12-16 所示。

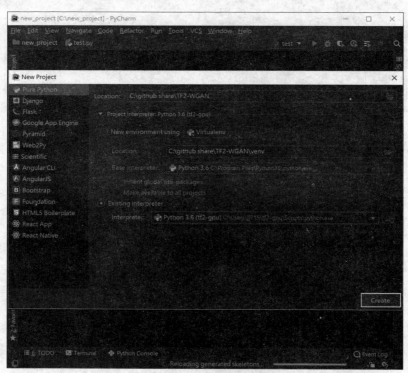

图 12-16　建立项目

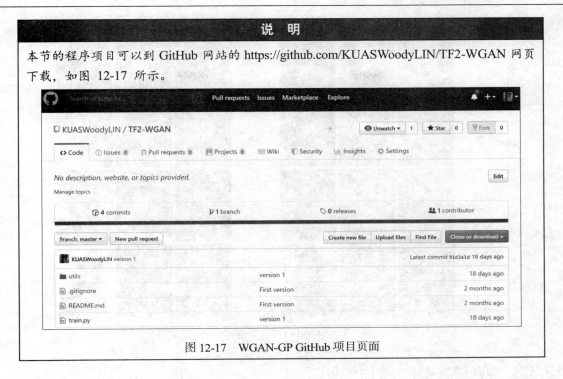

图 12-17　WGAN-GP GitHub 项目页面

12.3.2　数据集介绍

本节 WGAN-GP 的项目使用 Large-Scale CelebFaces Attributes Dataset（CelebA）[14]人脸数据集，这个数据集拥有 202 599 张图像，分别为训练数据 162 770 张图像、验证数据 19 962 张图像和测试数据 19 867 张图像，并且图像的大小为 218×178×3，如图 12-18 所示。可以直接通过 TensorFlow 来下载并加载这个数据集，程序代码如下：

```
import tensorflow_datasets as tfds

# 加载训练数据集
train_data, info = tfds.load("celeb_a", split= tfds.Split.TRAIN,
                        with_info=True)
# 加载验证数据集
valid_data = tfds.load("celeb_a", split= tfds.Split.VALIDATION)
# 加载测试数据集
test_data = tfds.load("celeb_a", split= tfds.Split.TEST)
```

图 12-18　Large-Scale CelebFaces Attributes Dataset 的部分图像

12.3.3　WGAN-GP 项目说明

WGAN-GP 项目的目录和文件

建立 WGAN-GP 项目的目录和 Python 文件，如图 12-19 所示。下面介绍需要建立的文件和目录。

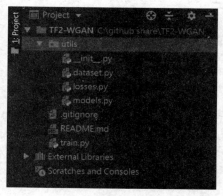

图 12-19　WGAN-GP 项目的目录和文件

- train.py

训练 WGAN-GP 模型的程序代码。

- utils

models.py：WGAN-GP 网络模型的程序代码。

losses.py：自定义损失函数的程序代码。
dataset.py：数据预处理的程序代码。

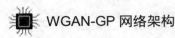 WGAN-GP 网络架构

首先，在 utils→models.py 的 Python 文件中编写 WGAN-GP 的网络架构，这里会建立生成器和判别器的网络模型，网络架构如图 12-20 所示。

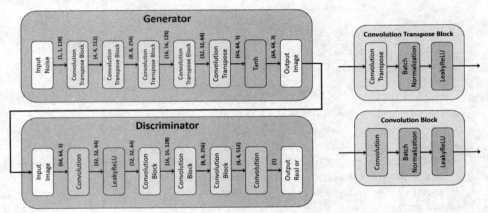

图 12-20　WGAN-GP 网络架构图

建立 WGAN-GP 网络模型的函数，该网络架构对照到图 12-20：

```python
def Generator(input_shape=(1, 1, 128), name='Generator'):
    inputs = keras.Input(shape=input_shape)
    
    # 1: Convolution Transpose Block1, 1x1 -> 4x4
    x = keras.layers.Conv2DTranspose(512, 4, strides=1, padding='valid',
                                     use_bias=False)(inputs)
    x = keras.layers.BatchNormalization()(x)
    x = keras.layers.LeakyReLU()(x)
    # 2: Convolution Transpose Block2, 4x4 -> 8x8
    x = keras.layers.Conv2DTranspose(256, 4, strides=2, padding='same',
                                     use_bias=False)(x)
    x = keras.layers.BatchNormalization()(x)
    x = keras.layers.LeakyReLU()(x)
    # 3: Convolution Transpose Block3, 8x8 -> 16x16
    x = keras.layers.Conv2DTranspose(128, 4, strides=2, padding='same',
                                     use_bias=False)(x)
    x = keras.layers.BatchNormalization()(x)
    x = keras.layers.LeakyReLU()(x)
    # 4: Convolution Transpose Block4, 16x16 -> 32x32
    x = keras.layers.Conv2DTranspose(64, 4, strides=2, padding='same',
                                     use_bias=False)(x)
    x = keras.layers.BatchNormalization()(x)
    x = keras.layers.LeakyReLU()(x)
    # 5: Convolution Transpose + Tanh, 32x32 -> 64x64
    x = keras.layers.Conv2DTranspose(3, 4, strides=2, padding='same',
```

```
                                             use_bias=False)(x)
        outputs = keras.layers.Activation('tanh')(x)
        return keras.Model(inputs=inputs, outputs=outputs, name=name)

    def Discriminator(input_shape=(64, 64, 3), name='Discriminator'):
        inputs = keras.Input(shape=input_shape)
        # 1: Convolution + LeakyReLU, 64x64 -> 32x32
        x = keras.layers.Conv2D(64, 4, strides=2, padding='same')(inputs)
        x = keras.layers.LeakyReLU()(x)
        # 2: Convolution Block1, 32x32 -> 16x16
        x = keras.layers.Conv2D(128, 4, strides=2, padding='same', use_bias=False)(x)
        x = keras.layers.BatchNormalization()(x)
        x = keras.layers.LeakyReLU()(x)
        # 3: Convolution Block2, 16x16 -> 8x8
        x = keras.layers.Conv2D(256, 4, strides=2, padding='same', use_bias=False)(x)
        x = keras.layers.BatchNormalization()(x)
        x = keras.layers.LeakyReLU()(x)
        # 4: Convolution Block3, 8x8 -> 4x4
        x = keras.layers.Conv2D(512, 4, strides=2, padding='same', use_bias=False)(x)
        x = keras.layers.BatchNormalization()(x)
        x = keras.layers.LeakyReLU()(x)
        # 5: Convolution, 4x4 -> 1x1
        outputs = keras.layers.Conv2D(1, 4, strides=1, padding='valid')(x)
        return keras.Model(inputs=inputs, outputs=outputs, name=name)
```

WGAN-GP 损失函数

定制生成器损失函数，这个损失函数的目的是让生成器生成的图像能使判别器判别为接近真实的图像，可以把程序代码编写在 utils→losses.py 文件中。

$$\text{Generator Loss} = -\frac{1}{N}\sum_{i=1}^{N} D(G(z^i))$$

N：训练数据量。
D：判别器。
G：生成器。
z：从一个分布中抽样作为生成器的输入。

定制生成器损失函数，把程序代码编写在 utils→losses.py 文件中：

```
def generator_loss(fake_logit):
    g_loss = - tf.reduce_mean(fake_logit)
    return g_loss
```

- 定制判别器损失函数，其中降低 real_loss 项可以让判别器判别真实图像更真；降低 fake_loss 项可以让判别器判别生成图像更假，而梯度惩罚项则可以让判别器满足 1-Lipschitz 函数，可以把程序代码编写在 utils→losses.py 文件中。

$$\text{Discriminator Loss} = \underbrace{-\frac{1}{N}\sum_{i=1}^{N}D(x^i)}_{\text{real\_loss}} + \underbrace{\frac{1}{N}\sum_{i=1}^{N}D(\hat{x}^i)}_{\text{fake\_loss}} + \lambda \times \text{Gradient Penalty}$$

N：训练数据量。

D：判别器。

x：从数据集中抽样的真实图像。

$\hat{x}$：从生成器得出的生成图像。

λ：梯度惩罚系数通常设置为 10。

```python
def discriminator_loss(real_logit, fake_logit):
    real_loss = - tf.reduce_mean(real_logit)
    fake_loss = tf.reduce_mean(fake_logit)
    return real_loss, fake_loss
```

$$\text{Gradient Penalty} = \frac{1}{N}\sum_{i=1}^{N}(\|\nabla_{\tilde{x}}D(\tilde{x})\|_2 - 1)^2$$

$$\tilde{x} = tx + (1-t)\hat{x}$$

N：训练数据量。

D：判别器。

$\tilde{x}$：从真实图像（x）和生成图像（$\hat{x}$）中抽样，将两个样本进行线性内插。

t：为 0~1 均匀分布的随机抽样数值。

```python
def gradient_penalty(discriminator, real_img, fake_img):
    def _interpolate(a, b):
        shape = [tf.shape(a)[0]] + [1] * (a.shape.ndims - 1)
        alpha = tf.random.uniform(shape=shape, minval=0., maxval=1.)
        inter = (alpha * a) + ((1 - alpha) * b)
        inter.set_shape(a.shape)
        return inter
    # 将生成图像与真实图像进行线性内插
    x_img = _interpolate(real_img, fake_img)
    with tf.GradientTape() as tape:
        # 确保 x_img 可以被 tape 追踪
        tape.watch(x_img)
        # 判别器会判别 x_img 的真假
        pred_logit = discriminator(x_img)
    # 计算梯度
    grad = tape.gradient(pred_logit, x_img)
    # 计算梯度的范数
    norm = tf.norm(tf.reshape(grad, [tf.shape(grad)[0], -1]), axis=1)
    # L2 正则化，希望损失值越接近 1 越好
    gp_loss = tf.reduce_mean((norm - 1.)**2)
    return gp_loss
```

数据预处理

将图像大小调整为 64×64，每个像素取值为 0~255，正则化为 -1~+1，可以把程序代码写在 utils→dataset.py 文件中。

```python
def parse_fn(dataset, input_size=(64, 64)):
    x = tf.cast(dataset['image'], tf.float32)
    crop_size = 108
    # 图像大小(218, 178, 3)
    h, w, _ = x.shape
    # 从图像中间裁剪(108, 108, 3)大小作为新的图像
    x = tf.image.crop_to_bounding_box(x, (h-crop_size)//2, (w-crop_size)//2,
                                      crop_size, crop_size)
    # 重新调整图像大小(108, 108, 3) ->(64, 64, 3)
    x = tf.image.resize(x, input_size)
    # 将图像像素值标准化到-1~+1，步骤：[0~255]/127.5→[0~2], [0~2]-1→[-1~1]
    x = x / 127.5 - 1
    return x
```

训练 WGAN-GP

这部分程序用于训练 WGAN-GP 的网络模型，可以把程序代码编写在 train.py 文件中。

Step 01 导入必要套件。

```python
import numpy as np
import tensorflow as tf
import tensorflow_datasets as tfds
from functools import partial
from utils.dataset import parse_fn
from utils.losses import generator_loss, discriminator_loss, gradient_penalty
from utils.models import Generator, Discriminator
```

Step 02 设置训练参数。

```python
# 批量大小（Batch）设置为 64
batch_size = 64
# learning rate 设置为 1×10⁻⁴
lr = 0.0001
# 输入生成器的域大小为 128 维
z_dim = 128
# 每训练 5 次判别器，训练 1 次生成器
n_dis = 5
# 设置梯度惩罚系数，通常设置为 10
gradient_penalty_weight = 10.0
```

Step 03 加载 CelebA 数据集，并设置数据集。

```python
# 设置读取到的数据集为训练数据、验证数据和测试数据
combine_split = tfds.Split.TRAIN + tfds.Split.VALIDATION + tfds.Split.TEST
# 加载数据集
```

```
train_data, info = tfds.load('celeb_a', split=combine_split, with_info=True)

# 自动调整模式
AUTOTUNE = tf.data.experimental.AUTOTUNE
# 打乱数据集
train_data = train_data.shuffle(1000)
# 加载预处理 parse_fn 函数，CPU 数量为自动调整模式
train_data = train_data.map(parse_fn, num_parallel_calls=AUTOTUNE)
# 设置批量大小为 64，如果最后一批数据小于 64，就舍弃该批数据
train_data = train_data.batch(batch_size, drop_remainder=True)
# 开启预取模式（缓存空间为自动调整模式）
train_data = train_data.prefetch(buffer_size=AUTOTUNE)
```

Step 04 建立生成器和判别器网络模型。

```
generator = Generator((1, 1, z_dim))
discriminator = Discriminator((64, 64, 3))
```

Step 05 设置生成器和判别器的训练优化器。

```
g_optimizer = tf.keras.optimizers.Adam(lr, beta_1=0.5, beta_2=0.9)
d_optimizer = tf.keras.optimizers.Adam(lr, beta_1=0.5, beta_2=0.9)
```

Step 06 建立生成器训练函数。

函数会去计算生成器的损失函数，并计算梯度，更新生成器网络权重。

```
@tf.function
def train_generator():
    with tf.GradientTape() as tape:
        # 从正态分布中产生 128 维的随机向量作为生成器
        random_vector = tf.random.normal(shape=(batch_size, 1, 1, z_dim))
        # 从生成器中生成假的图像
        fake_img = generator(random_vector, training=True)
        # 使用判别器评估生成图像
        fake_logit = discriminator(fake_img, training=True)
        # 计算生成器损失
        g_loss = generator_loss(fake_logit)
    # 计算梯度
    gradients = tape.gradient(g_loss, generator.trainable_variables)
    # 更新生成器权重
    g_optimizer.apply_gradients(zip(gradients, generator
                                    .trainable_variables))
    return g_loss
```

Step 07 建立判别器训练函数。

此函数会计算判别器的损失函数，并计算梯度，更新判别器网络权重。

```
@tf.function
def train_discriminator(real_img):
    with tf.GradientTape() as t:
```

```python
                # 从高斯分布中产生128维的随机向量作为生成器
                random_vector = tf.random.normal(shape=(batch_size, 1, 1, z_dim))
                # 从生成器中生成假的图像
                fake_img = generator(random_vector, training=True)
                # 使用判别器评估真实图像
                real_logit = discriminator(real_img, training=True)
                # 使用判别器评估生成图像
                fake_logit = discriminator(fake_img, training=True)
                # 计算真实图像和生成图像的损失
                real_loss, fake_loss = discriminator_loss(real_logit, fake_logit)
                # 计算梯度惩罚
                gp_loss = gradient_penalty(partial(discriminator,training=True),
                                           real_img,fake_img)
                # 计算判别器损失
                d_loss = (real_loss + fake_loss) + gp_loss * gradient_penalty_weight
        # 计算梯度
        D_grad = t.gradient(d_loss, discriminator.trainable_variables)
        # 更新判别器权重
        d_optimizer.apply_gradients(zip(D_grad, discriminator.trainable_variables))
        return real_loss + fake_loss, gp_loss
```

说 明

tf.function 介绍

TensorFlow 2 默认为 Eager Execution 动态图模式，这个模式一旦执行，就会立刻返回数值，提供了更灵活的使用，但可能会牺牲一些性能。为了使性能优化，TensorFlow 2 推出了 @tf.function 修饰器，使用这个修饰器的函数，会通过一个名为 AutoGraph 的工具将程序代码转为静态计算图（例如 while->tf.while、if->tf.cond 等）。

更多详细介绍可到 TensorFlow 官方网站查阅：https://www.tensorflow.org/versions/r2.0/api_docs/python/tf/function。

Step 08 生成100张图像。

将生成的100图像放入10×10的大数组中显示。

```python
def combine_images(images, col=10, row=10):
    # 为了让生成图像正常显示，将图像从-1~+1 缩放到 0~1
    images = (images + 1) / 2
    # 将 TensorFlow 格式转换成 Numpy 格式
    images = images.numpy()
    # 取得生成图像的形状，shape=(batch size, height, width, channel)
    b, h, w, _ = images.shape
    # 建立10×10的大数组存储100张图像
    images_combine = np.zeros(shape=(h*col, w*row, 3))
    # 将100张图像放入10×10的大数组中
    for y in range(col):
        for x in range(row):
            images_combine[y*h:(y+1)*h, x*w:(x+1)*w] = images[x+y*row]
    return images_combine
```

Step 09 WGAN-GP 训练程序。

这部分为 WGAN-GP 的训练主体，训练程序每训练 5 次判别器，就会训练 1 次生成器。

```python
def train_wgan():
    # 建立存储生成器模型的目录
    log_dirs = 'logs_wgan'
    model_dir = log_dirs + '/models/'
    os.makedirs(model_dir, exist_ok=True)
    # 建立 TensorBoard 日志
    summary_writer = tf.summary.create_file_writer(log_dirs)
    
    # 从正态分布中产生一组固定的随机向量（作为验证用）
    sample_random_vector = tf.random.normal((100, 1, 1, z_dim))
    # 总共训练 25 个 epoch
    for epoch in range(25):
        # 读取训练数据（真实图像）
        for step, real_img in enumerate(train_data):
            # 训练判别器
            d_loss, gp = train_discriminator(real_img)
            # 把判别器的损失值存储到 TensorBoard 日志
            with summary_writer.as_default():
                tf.summary.scalar('discriminator_loss', d_loss, 
                                  d_optimizer.iterations)
                tf.summary.scalar('gradient_penalty', gp, 
                                  d_optimizer.iterations)
            
            # 每训练 5 次判别器，执行 1 次生成器训练
            if d_optimizer.iterations.numpy() % n_dis == 0:
                # 训练生成器
                g_loss = train_generator()
                # 把生成器的损失值存储到 TensorBoard 日志
                with summary_writer.as_default():
                    tf.summary.scalar('generator_loss', g_loss, 
                                      g_optimizer.iterations)
                # 显示当前生成器、判别器和梯度惩罚的损失值
                print('G Loss: {:.2f}\tD loss: {:.2f}\tGP Loss {:.2f}'.format(
                    g_loss, d_loss, gp))
                
                # 生成器每训练 100 次，会产生 100 张图像并存储到 TensorBoard
                if g_optimizer.iterations.numpy() % 100 == 0:
                    # 从生成器中产生 100 张生成图像
                    x_fake = generator(sample_random_vector, training=False)
                    # 将生成的 100 图像放入 10×10 的大数组中显示
                    save_img = combine_images(x_fake)
                    # 把 100 张生成图像存储到 TensorBoard 日志
                    with summary_writer.as_default():
                        tf.summary.image(dataset, [save_img], 
                                         step=g_optimizer.iterations)
        # 每一个 epoch 存储一次生成器模型权重
        if epoch != 0:
```

```
generator.save_weights(model_dir+"generator-epochs-{}.h5"
                       .format(epoch))
```

 用 TensorBoard 观察训练

最后可以通过 TensorBoard 观察生成器训练的预测变化，如图 12-21 所示，以及观察生成器损失、判别器损失和梯度惩罚损失的曲线图，如图 12-22 和图 12-23 所示。

图 12-21　用 TensorBoard 观察生成器训练的预测变化

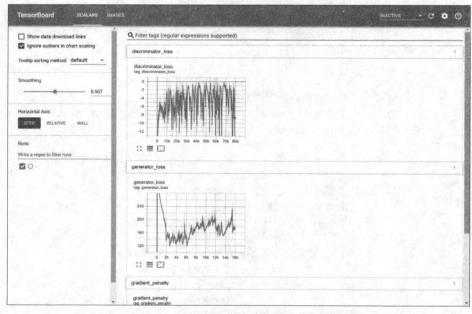

图 12-22　用 TensorBoard 观察生成器损失和判别器损失的曲线图

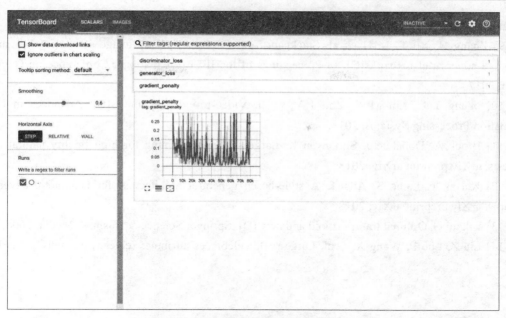

图 12-23　梯度惩罚损失的曲线图

启动 TensorBoard（命令行）：

```
tensorboard --logdir logs-wgan
```

12.4　参考文献

[1] Goodfellow I, Pouget-Abadie J, Mirza M, et al. Generative adversarial nets[C]. Advances in Neural Information Processing Systems, 2014, 2672-2680.

[2] Radford A, Metz L, Chintala S. Unsupervised representation learning with deep convolutional generative adversarial networks[C]. In International Conference on Learning Representations, 2016, 1-16.

[3] Salimans T, Goodfellow I, Zaremba W, et al. Improved techniques for training gans[C]. Advances in Neural Information Processing Systems, 2016, 2226-2234.

[4] Lin Z, Khetan A, Fanti G, et al. PacGAN: The power of two samples in generative adversarial networks [C]. Advances in Neural Information Processing Systems, 2018, 1503-1512.

[5] Arjovsky M, Chintala S, Bottou L. Wasserstein GAN [C]. In Proc. 34th International Conference on Machine Learning, 2017, 214-223.

[6] Gulrajani I, Ahmed F, Arjovsky M, et al. Improved training of wasserstein GANS[C]. Advances in Neural Information Processing Systems, 2017, 5767-5777.

[7] Zhu J-Y, Park T, Isola P, et al. Unpaired image-to-image translation using cycle-consistent adversarial networks[C]. In International Conference on Computer Vision, 2017, 2223-2232.

[8] Karras T, Aila T, Laine S, et al. Progressive growing of GANs for improved quality, stability,

and variation [C]. In International Conference on Learning Representations, 2018, 1-26.

[9] Zhang H, Xu T, Li H, et al. Stackgan: Text to photo-realistic image synthesis with stacked generative adversarial networks[C]. In Proceedings of the IEEE International Conference on Computer Vision, 2017.

[10] Wang T C, Liu M Y, Zhu J Y, et al. Video-to-video synthesis[C]. Advances in Neural Information Processing Systems, 2018.

[11] Brock A, Donahue J, Simonyan K. Large scale gan training for high fidelity natural image synthesis. arXiv preprint arXiv, 2018.

[12] Karras T, Laine S, Aila T. A style-based generator architecture for generative adversarial networks. arXiv preprint arXiv, 2018.

[13] Villani C. Optimal transport: old and new [D]. Springer Science & Business Media, 2008, 338.

[14] Liu Z, Luo P, Wang X, et al. Large-scale celebfaces attributes (celeba) dataset[C]. Retrieved, 2018.